Industrial Poultry Keeping

Industrial Poultry Keeping

Edward Brown

2025
Agri Horti Press
New Delhi-110 084

ISBN: 978-93-49690-87-5

Published By

Agri Horti Press

New Delhi

E-mail: agrihortipress@gmail.com

Dark (or Coloured) Dorkings.

PREFACE TO THE FOURTH EDITION

THE great success which attended my larger work on 'Poultry-Keeping as an Industry for Farmers and Cottagers,' and the movement towards a development of poultry-culture as one of our rural industries, led to the preparation of this work for cottagers and small farmers, and its success has shown that it met this need. In it no attempt is made to enter at all fully into the many details given in the work named above, but no point is omitted which can be of service to those who seek to improve their own position by following a pursuit only too much neglected in the past. The purely fancy aspect is not dealt with at all, for the author's desire has been to provide a handy guide to the first steps of poultry-keepers.

So far as cottagers are concerned, my own hope is to see the day when every rural labourer is not only a worker for someone else, but a producer himself. He will be a stronger man when that ideal is reached. More contented in mind, better able to meet the troubles of life, with an object before him other than merely animal, and relieved from the monotony of existence too often found in country districts, he will be a nobler citizen and a truer man. The efforts of the past decade, both as to greater opportunities and increased knowledge, and especially the technical instruction work of County Councils, are apparent by the larger number of fowls kept in all parts of the country. But there is much yet to be done. I am sorry to say that in some districts land-owners and farmers still repress the keeping of poultry by their workers. The reason adduced is that they fear, if permitted, food would be stolen to feed the poultry. Perhaps it might be so in a few instances, but that it would be at all general, or even sufficiently common to warrant repressive measures, I cannot for one moment believe. The experience of those districts where poultry-keeping is not so discouraged is sufficient evidence in favour of greater trust being shown. Suspicion breeds crime all the world over.

Since this work was first published, many changes and developments have taken place and solid progress been made. These changes and developments are, to some extent, indicated by the present edition, which has been carefully revised and some fresh matter introduced. Increased production necessitates new systems of marketing, and I have therefore added a chapter dealing with this important question, which must be met by methods similar to those adopted in our colonies and foreign countries.

EDWARD BROWN.

Contents

ILLUSTRATIONS

PLATES

BLOCKS

Industrial Poultry-Keeping

CHAPTER I

INTRODUCTION

Small Industries.—The trend of all recent developments, so far as our rural population is concerned, is undoubtedly rather towards the encouragement of small than of great things. This fact is apparent on every side. What force there has been in the pressure of circumstances during the last three decades is accentuated by the great movement in the direction of Technical and Secondary Education, which is doing so much to elevate the working people of this country. Too long has there been a neglect of teaching those details of daily occupation which are essentially connected with the position of our country as a whole, and every individual member of the community whose living depends upon mental or physical labour. That neglect is now at an end, and it is a cause for universal satisfaction that the County Councils, with one or two exceptions, have utilized the grants placed in their hands through the Local Taxation Act, 1890, to promote those branches of education already referred to. By means of the grants these public bodies are able to do that which was scarcely possible before, and the importance of what are undoubtedly minor branches of industry is receiving a recognition we had scarcely thought possible at one time. Hence there is an upheaval at work, and an opportunity of improving the position of our people generally, which has now for the first time been presented. Much naturally depends upon the way in which advantage is taken of the teaching offered. Lethargy there has been, but in many districts the interest awakened in the possibilities of poultry-keeping has been very great.

Will Poultry Pay?—We know it has been a common dictum that 'poultry will not pay,' due to the failure of pretentious attempts at establishing poultry farms pure and simple. These have been from time to time heralded with loud trumpets as a means of securing wealth, and great has been their downfall. The reasons for these failures are not far to seek, and need not be dwelt upon here. But because such enterprises have brought disaster upon their promoters is surely no proof that the keeping of poultry by farmers as part of their farm stock, or in a small way by cottagers, is unprofitable. In fact, there is an abundance of instances in which the reverse is the case, not merely in the great poultry and egg-producing districts abroad, but also in Devon and Cornwall, in Sussex and Surrey, in Buckinghamshire and Norfolk, in Cumberland and Lincolnshire, and in more isolated cases throughout the entire country. Many farmers now find their poultry an important element in securing a good balance-sheet, and numerous cottagers add considerably to their incomes by this pursuit. In Ireland it is so to an even greater extent. Since 1888, when I was commissioned by the *Weekly Freeman* of Dublin to undertake an exhaustive inquiry into the poultry industry of the Green Isle, a vast progress has been made, and the number of fowls kept increased by several millions.

Britain's Egg and Poultry Bill.—Dealing specially,

however, with Great Britain, what do we find? That nearly all the great centres of population are almost entirely dependent upon foreign eggs during several months of the year, and upon Irish at other seasons, the home supplies often being unreliable and erratic. There are rural districts where home eggs do not even supply the needs of their residents. Even in market towns, surrounded by pastoral districts, with no great industrial population near, foreign and Irish eggs have to be imported, and I do not think that any other counties than Devon and Cornwall actually supply their own needs in this respect. Hence we can understand the great and growing imports from abroad. In my work on "Poultry-Keeping as an Industry for Farmers and Cottagers," I give the tables of imports in detail, and need not repeat them here, but in order to show the importance of this subject I may mention that in thirty-five years the imports have increased sevenfold, as the following figures will show:

	£
1864	835,028
1880	2,335,451
1890	3,926,660
1899	5,829,686

In addition to foreign imports there is the Irish supply, which I calculate at about £2,000,000, bringing up the grand total to nearly £8,000,000 sterling per annum—no small amount. London alone spends about £1,500,000 in foreign eggs and poultry, and it is within the mark to say that Yorkshire spends on Irish and foreign eggs and poultry £820,000, and Lancashire £850,000, every year, with other manufacturing districts in proportion. In face of these facts, and with steadily increasing imports, the question assumes an importance which has never yet been fully recognised in the minds of many who have every opportunity for poultry-keeping, but have not thought it worthy of their consideration. Whilst we would encourage fruit-keeping to the full, because it is a profitable pursuit, and we desire to see this article of food more fully used than ever it has been, it is only right to point out that the relative position is in favour of eggs and poultry if we take those fruits which can be produced in this country ; therefore it is only reasonable to ask that the attention given to one shall also be extended to the other.

Industrial Poultry - keeping.—The object of our present work is to provide the best and latest information, derived from experience both at home and abroad, upon the subject of poultry-keeping, alike for the small farmer who desires to add this to his other operations, and the cottager who can only keep a few fowls, but to whom those few are of equal import as are the many to someone else. This information, it is intended, shall be concise and practical. Those who desire to go into the matter more fully are recommended to obtain the larger book, 'Poultry-Keeping as an Industry for Farmers and Cottagers,' wherein will be found all facts obtainable as to the best methods adopted both at home and abroad, with details which are impossible in a work of the present dimensions. This may be taken as more suitable for those who intend to limit their operations, or as a guide to the first steps in industrial poultry-keeping. The experience gained during the past few years has proved that there are possibilities in poultry-keeping which were not realized before. Whilst what is sometimes called poultry-farming, though we do not agree to that term being so restricted, namely, devoting land to poultry alone, and depending entirely upon poultry produce for an adequate return, has not succeeded, modifications have proved successful. Many cases could be given in proof. Usually, however, the sale of breeding stock and of eggs for hatching secure enhanced profits. But later developments have extended our knowledge on this subject, and American experience has shown that operations carried out under proper conditions are not so restricted as we supposed. The dangers of this branch are great. These are referred to in Chapter IV., and, if avoided, a new industry will be open to our rural population. Farmers of all grades, to attain success, should give the same attention to the fowls as they devote to larger stock, and associate them in the rotation of work. Cottagers and those who have only a small area of land must understand the principles underlying cultivation as well as the breeding and management of live-stock.

Poultry-keeping Abroad.—When we look at other countries, it is at once seen that the great mass of poultry produce sent by them is not due to extensive poultry-farms, but to the multitudinous small farmers and cottagers found there. Nearly all of these are poultry-keepers. The number of birds bred by each may not be great, but each believes in the business as a profitable one, and their combination has provided in a wonderful manner for our needs. It has been my good fortune to visit most of the countries of Western and Northern Europe in order to study the methods adopted, and the conclusion arrived at is, provided that the most advanced methods of management and marketing are adopted, our own people can compete with foreign producers and secure the best and most paying part of the trade. But this will only be by taking advantage of the nearness to markets and sending forward eggs and poultry in the best condition, prepared in the manner desired by retailers and consumers. To compete with Russian or Italian or even Danish eggs would be to sacrifice everything. As it may be hoped that erelong it will be possible for every cottager to have more or less land for his own cultivation, and that farms will return to their former more modest dimensions, we look for a development of poultry-keeping and other rural industries of a like nature.

CHAPTER II

COMMENCING

Begin quietly.—Poultry-keeping must be carried out upon the same lines as any other business, if success is to be attained. There can be no doubt that the best way to secure success in any undertaking is to commence in a modest way and gradually increase as opportunity offers. If the history of our great mercantile firms were written, it would be found that those which are the strongest had very small beginnings, and that of those which commenced with flash and noise very many have come down with a great crash. We do not propose to discuss social problems in this place, but may be permitted to say that we have no faith in the modern system of business which sinks the individual in the mass. A large establishment, which has been gradually developed, will probably be carefully and economically worked; but one commenced upon a large scale is seldom so. And the same thing is found in poultry-keeping. Therefore, let the commencement be small, gain experience as you go on, then quietly enlarge as capability to manage and demand for the produce develop; and, although the profit for the first year or two may not be so great as might have been wished, yet this system will pay best in the long-run. One of the features which I have strongly advocated is that we must chiefly look for profitable poultry-keeping to

WHITE DORKINGS.

Ludlow
1892

farmers and cottagers, who make it part of their farm operations, and not the first or only object in view. I shall show later on in what direction the poultry business can be made profitable as a separate industry, but it is not in raising the chicks or keeping vast numbers of hens for egg production alone. Frequently those who have succeeded best were such as did not depend upon the poultry as a means of livelihood—at any rate, during the first two or three years.

How to Begin.—Beginning by keeping poultry on a large scale is more attractive than starting quietly, but we should certainly advise anyone establishing a poultry-farm to commence by purchasing ten to twenty hens and one or two cocks. If this be done during the summer and autumn, they can be obtained at reasonable prices, and by the following spring all be ready for commencing operations in good earnest. From each hen—that is, if of a good laying variety and the season be an average one—forty or fifty chickens may reasonably be expected if all the eggs laid be set. Thus, from twenty hens nearly 1,000 chicks can be secured, and as about half of these will probably be cockerels, and a percentage must be allowed for deformed birds and deaths, the laying stock at the commencement of the following winter should be about 400 hens. This number can be increased as desired, but it is not advisable to augment the stock too rapidly. The time spent before the laying hens are ready is by no means wasted, as much can be done in preparing the ground, building houses, and acquiring experience which will be found of value afterwards. It is of no use leaving the building of the houses until they are required, and often have we known a whole season wasted by postponing this important work until the spring, thus making hatching operations too late to be of real service. The time to spare can be utilized in many ways, as already indicated; and if the ground is not already supplied with trees or bushes, the runs may have a goodly number planted, not only as a protection to the fowls in sunny or wet weather, but, if fruit-trees are used, as a source of profit.

False Steps.—We know there are many would-be poultry-keepers who prefer purchasing a large number of birds, and thus getting to work without delay; but the first cost must be very much greater than if the plan suggested be adopted. There is also the risk of buying unhealthy birds, and the certainty of having to take old ones and of breeds not suited to the end in view. We once knew an instance of an attempt to commence a poultry-farm, where the owner went into the market and bought about 2,000 hens irrespective of age or breed. It was, after a few months of apparent success, a dismal failure, as was anticipated by all who knew anything of poultry-keeping. Young hens are wanted for profit, and young hens cannot be bought in large numbers. Old ones are dear almost at a gift, and therefore such as are required must be bred. Further, a good laying strain is only secured by selection, and this must be made by the individual breeder.

Cottage Poultry-keeping.—In connection with nearly all cottages there is some outbuilding which can easily be turned into a poultry-house. Upon this point we speak in a succeeding chapter, and need only mention it here as showing that a start can be made at once. The simplest way is to buy a cock and half a dozen hens, taking care that they are of a right variety and young birds. In this way the return will probably be immediate, for, as a rule, they will commence to lay forthwith. The cheapest time of year to purchase is in the autumn. Then nearly all breeders have a quantity of surplus stock, which they are glad to dispose of at reasonable prices; whereas if purchasers wait until after Christmas, when these have been cleared out, two or three times as much will have to be paid for the same birds, with a more restricted choice. A cheaper plan is to buy eggs in the spring, and depend upon the chickens hatched from them; and if good eggs can be obtained this is a satisfactory method, though better for improvement of present stocks than in making a commencement, as it will only then be requisite to purchase a cock bird in the autumn for the next season's breeding. Of course, immediate returns cannot be secured in this manner, but where the means are limited it may be preferred. If more cockerels are obtained than the owner requires, they can be turned into money in about three months from the time of hatching. We should say that it is a wiser plan to purchase eggs as near home as possible rather than send long distances to unknown persons. There is, unfortunately, a large amount of swindling in the selling of eggs, but this has been in some measure checked by the precautions taken in connection with poultry journals for the protection of their readers, and many of the best breeders give a guarantee as to fertility. Eggs vary in price from 3s. 6d. to 21s. per dozen, and capital settings can be obtained at about 5s. to 6s. each.

Pure *v*. Cross Breeds.—It may be desirable to here point out the greater advantages from the keeping of pure-bred fowls than cross-breeds or mongrels. And these remarks apply equally to farmers and cottagers. In some districts there is a strong feeling in favour of the so-called 'barn-door' fowl, whatever that may be. This term is usually applied to mongrels of every shape and kind, and they vary with every county, if not parish. First crosses between two pure-breds may be used with advantage to obtain desired results; but when intermixing is carried further there is a speedy deterioration. The advantage of pure-bred stock is that we can depend upon it, knowing that it has been bred upon continuous lines; whilst with mongrels the influences at work are unknown quantities, and in many cases these are in direct opposition to each other. The experience of the past has shown us that

Britain owes her pre-eminence as a great stock-breeding centre to rigid adherence to type in horses, cattle, and other domesticated animals. The same applies equally to poultry. We therefore strongly urge that all denominations of poultry-keepers should select some pure-bred fowl and adhere to it. If there is sufficient space at command, it may then be wise to produce cross-bred fowls, either as layers or for the table; but to do this it is necessary to have two pure races, which must be kept as the basis. There is yet another advantage to be secured by adopting this suggestion, namely, that the surplus stock are so much more valuable, and eggs sold for hatching will command a higher price. A cross-bred fowl may bring as much when dead as one of the purest strains, and the eggs from one realize the same as from the other if marketed in the regular manner; but buyers will give double as much for a pure cock or hen for breeding, or for eggs from pure fowls, as for or from cross-breds. To cottagers this is not so important as in the case of farmers, as they have not the same opportunities for selling their stock, but even with them it is worthy of consideration. Economically nearly all pure-bred fowls are as good as any cross ever made, and they are certainly more dependable.

Another Method of Improvement.—When there is a stock of young hens on hand, which there is a disinclination to sell or kill, much can be done by introducing one or more pure-bred cockerels of whatever variety is thought most suitable. This is the plan which has been followed with great success in Ireland. Where adopted, a sufficient number of the half-bred pullets should be selected and bred from the second year, when they must be mated with another cockerel of the same variety, but unrelated, and the same process repeated the third year. In this way, by a very small expenditure, the stock can be entirely rejuvenated. In fact, if three persons in the same district adopted this plan, they would only need to buy one male bird each, and exchange them the following years. The points to be carefully observed are, (1) that there shall be no in-breeding; (2) that only pure cocks be employed; and (3) that male birds of the same variety shall be used each year.

Accounts.—All poultry-keepers ought to keep accounts of their operations, not merely for the satisfaction of knowing whether there has been a profit or loss, though that is very desirable, but also as an inducement to economy. There are several poultry account-books sold which may be used, or a plain book may be ruled to serve the purpose. At the beginning of operations or of the year, all the birds and appliances should be valued and recorded. Next, all items of receipt and expenditure should be set down regularly and methodically; then the number of eggs laid, and how they are disposed of, whether sold, consumed in the house, used for sitting, or for the chickens; and finally a hatching record, showing the dates and number of eggs set, how many hatched, and the number reared. At the year's end there will have to be another valuation, and the debit side will show the valuation at the beginning of the year and the expenditure; whilst the credit side will show the receipts, the allowance for eggs and chickens consumed, and the final valuation. The balance between one side and the other will represent the profit or loss, whichever has resulted.

CHAPTER III

BREEDS OF DOMESTIC POULTRY

There are in all some forty breeds of domestic poultry, but more than half of these do not call for reference in a work of this kind. They are either purely fancy fowls little known, or their economic merits pale before the varieties referred to below. Those who care for the external characteristics more than for internal qualities, and look at poultry-keeping from a fancy rather than a practical point of view, should consult 'Pleasurable Poultry-Keeping,' in which will be found descriptions of the leading varieties.

Comparison of Breeds.—The question is often asked, Which is the best breed of fowls? To such a question it is impossible to give an answer, for it is, in the first place, necessary to determine a definition of the word 'best.' What would make the breed 'best' to one man might not be any recommendation to another. The breeder who desires first of all high table quality would not be content with the great laying powers but moderate flesh of the non-sitting varieties, and thus it is impossible off-hand to give a reply. The only way is to classify as far as possible the breeds under the qualities for which they are most noted, and then the reader will have some notion as to which is best for his purpose. In each section, as far as possible, the varieties are placed in the order of their respective qualities.

Laying (or Non-Sitting) Varieties.

Hamburghs.	Andalusians.
Campines.	Redcap.
Minorcas.	Houdan.
Leghorns.	Scotch Greys.
Anconas.	Barbezieux.

Table Varieties.

Dorkings.	Crevecœur.
Indian Game.	Coucou de Malines.
Game.	Courtes Pattes.
La Bresse.	Du Mans.
La Fleche.	Faverolle.

General Purpose Varieties.

Plymouth Rocks.	Orpingtons.
Wyandottes.	Brahmas.
Langshans.	

These divisions will show the best arrangement of the different varieties of fowls, and may be taken as the only classification that is at all practicable, approaching the matter from an economic aspect. Some of these breeds, of course, might come under the other heads, such, for instance, as Houdans, which make capital table birds, but on the whole experience has shown that the above arrangement is as nearly correct as can be.

Description of Breeds.—It is not necessary that we should go into very great detail as to the individual characteristics of different varieties of poultry, but our purpose will be served if we indicate the leading points (quoting the same from 'Poultry-Keeping as an Industry for Farmers and Cottagers'). For facility of reference we give these in alphabetical order, omitting such as are of no practical use to the poultry-keeper who is seeking to produce either eggs or poultry for market.

ANCONAS.—A mottled-plumaged variety, nearly resembling Leghorns; excellent layers; medium-sized body; very hardy and precocious.

ANDALUSIANS.—A member of the Mediterranean family, which has been described as having a smallish-sized body, placed upon legs of a good length; the neck rather long, with a fine head; a large upright single comb in the cock; in the hen also large, but falling over on one side; and the cocks have large sickle-shaped tails. Andalusians are clean-legged; colour slate, except on the cock's neck and back, where it is dark purple, nearly black.

Brahmas.—Large, full-feathered birds, feathers extending down the legs and feet; small, neat heads, with pea combs; two varieties—darks and lights.

Campines.—The great egg-producing breed of Belgium; resembling our pencilled Hamburghs, except that they have single combs; remarkable layers; small-sized bodies; two colours—gold and silver.

Coucou de Malines.—A large, upstanding breed, quick in growth, and fattening well; largely of Brahma type; cuckoo plumage; hardy.

Crevecœur.—A French breed of great size; great breast development; carries a medium-sized crest of feathers; legs black, but clean, *i.e.*, not feathered; plumage entirely black.

Dorkings.—Body large and deep, and when viewed sideways almost forming a square, breast well forward; neck short and head medium; legs clean, white in colour, and carrying a fifth toe; four varieties—darks, silver-greys, whites, and cuckoos.

Faverolle.—New variety introduced in Normandy; made from admixture of Houdan, Brahma, and Dorking blood; good layers and fatten well; slightly feathered in legs; single-combed, but uncertain in colour as yet; very hardy.

Game.—What are known as modern game are not so good as the old-fashioned fighting type, which are close, compact, and hard-feathered; bodies medium in size, broad in breast; legs clean, and rather long; several varieties, of which the white-legged are preferred for table purposes. The old English type of game is best for profit.

Hamburghs.—Small-sized, well-shaped bodies on longish legs; large sickle tail; full hackle, with neat

Indian Game.

head, and rose comb; five varieties—blacks, gold-spangled, silver-spangled, gold-pencilled, and silver-pencilled, all very rich in colour; save the blacks, they lay too small eggs to be of marketable value.

HOUDAN.—Of French origin; large eye, broad and massive; clean legs, pale in colour, carrying a fifth toe; head crested; plumage, mottled black and white.

INDIAN GAME.—Large, somewhat heavily-boned fowls, carrying a large amount of flesh on the breast; partridge in plumage, which is very rich in colour; clean, yellow legs; long neck and legs; small tail.

LA FLECHE.—Large French fowls; massive in body, deep in breast; clean black legs; neat head, with forked comb; plumage entirely black.

LANGSHANS.—Large Chinese fowls, on rather long legs; full tails, carried high; slightly feathered on the legs; comb single; plumage entirely black.

LEGHORNS.—Of the Mediterranean type; active fowls of great precociousness; legs clean and yellow; several varieties, white and brown being the oldest and best known; other colours are cuckoo, black, pile, duckwing, and buff.

MINORCAS.—Also of the Mediterranean family; two varieties, namely, black and white, but the latter seldom seen; one of the most valuable breeds we possess as egg-layers.

ORPINGTONS.—Really clean-legged Langshans, which they resemble, except that the legs are not feathered. A buff Orpington has lately been introduced, really a refined variety of the Lincolnshire Buff, found in the East of England.

PLYMOUTH ROCKS.—A breed of American composition; large in body; rather big in bone; clean yellow legs; four varieties—barred (cuckoo), whites, blacks, and buffs, the first-named being most popular.

REDCAP.—Original type of the gold-spangled Hamburgh; similar in colour, not so even in markings; very large comb; hardy, and most prolific layers.

SCOTCH GREY.—Most nearly like the Dorking in shape, and has white or speckled legs; plumage black and white.

WYANDOTTES.—Another breed of American production, and one of the most recent; comb rose; legs clean and yellow; large-sized body; plumage laced; in four colours or varieties—silvers, golds, whites, and buffs.

Ducks.—There are four leading varieties of ducks for economic purposes, though in this species is to be found a large number of ornamental or fancy breeds. Dividing these in regard to their chief qualities, we find them to be: For early maturity: Aylesbury. For large size: Rouen. For egg production: Pekin and Indian runner.

Their leading characteristics may be described as follows:

AYLESBURY.—Heavy in body, short in leg, and of a boat shape; bill long, and of a flesh colour; legs deep orange in colour; an excellent layer; very hardy; very rich in flesh; plumage entirely white.

ROUEN.—Similar in shape to the Aylesburys; grows to a greater size, but more slowly; preferred for the summer and autumn trade; plumage similar to that of the wild duck; a good layer, and fine in quality of flesh.

PEKIN.—Of Chinese origin; smaller in body than either of those named; not fleshy; more upright in carriage; plumage with a canary tinge.

INDIAN RUNNER.—A variety which has attained a considerable measure of popularity. It is small in size, but very delicate in flesh, and a wonderful layer.

Geese.—Here we find two principal varieties; the Toulouse is the larger, but not so fine in flesh. For early maturity: Embden. For large size: Toulouse.

Their leading characteristics are as follows:

EMBDEN.—Erect in shape for geese; pure white in plumage; flesh-coloured bills and orange-coloured legs and feet; quick in growth; suitable for Michaelmas trade.

TOULOUSE.—Head, neck, back, and thighs a dark-shaded brown-grey, with outer edges of each feather shaded a light grey; breast the same colour, but lighter between the legs, and white underneath; tail white, with band across centre of top; wing-flights grey; bill and feet dark orange; very large, but slow in growth.

Turkeys.—Of these valuable fowls there are three leading varieties bred for market. In this case there is a great demand for big fleshy specimens at Christmas, which realize a much higher price *pro rata* than do smaller birds, even though the latter are usually finer in quality of flesh.

For quality of flesh: Norfolk black.
For size and quantity of flesh: Cambridge bronze and American bronze.

The leading characteristics are as follows:

NORFOLK BLACK.—Medium size; very fine in flesh, which is of superior flavour; rather delicate; entirely black in plumage.

CAMBRIDGE BRONZE.—Upright carriage and commanding appearance; full, well-fleshed breast; full tail; grows well on suitable places; fairly hardy; plumage black shaded with bronze.

AMERICAN BRONZE.—Long, deep body, mounted on long and stout legs; broad and full breast; plumage black on parts brilliantly shaded with bronze, and bright bronze elsewhere; grows to enormous size; hardiest of all turkeys.

Choice of Breeds.—Another important question is that of the selection of stock, but this will largely depend upon what is most in demand. Chickens if bred and put upon the market are very profitable, but egg production offers, perhaps, the most regular field of operations, more especially for cottagers. There is no reason, however, why both should not enter into the work of a farm. A most important consideration is the soil upon which the fowls are to be kept. On light or medium lands either egg production or chicken-rearing may be carried out successfully, but on clay or heavy land eggs should be made the chief object, and a yellow-legged race selected, as they thrive better thereon. Let us, however, urge that proper stock be selected, that pure breeds be used for breeding crosses as layers and table-birds, but do not breed from the crosses, as it must ever be borne in mind that the progeny from cross-bred birds soon deteriorates. This is of great importance. 'Well begun is half done' is true of poultry-rearing as well as of every other department of life.

CHAPTER IV

HOUSES AND RUNS

The question of housing is most important, and must be settled in a great measure according to local circumstances. But on an ordinary farm the best plan is to use movable houses, dividing the flocks as much as possible, and giving fresh ground whenever it is thought necessary. If near an orchard or plantation, all the better, and the fowls will appreciate such an advantage to the full. Not more than fifty birds must be kept in one house and run, whether in confinement or at liberty; but it is found that twenty-five will lay a greater average number of eggs than if fifty are in a house relatively larger. If movable houses cannot be used—but there are few farms where this is so—and the fowls must be kept to a limited space, which is sometimes the case in hunting and densely-populated districts, it is better to build houses on the plan shown in Fig. 1, which provides for roosting-place and covered run. Internal arrangements, floors, etc., we deal with further on. Sometimes it is desirable to have a few confined runs for breeding pens; but the farmer is well advised if he avoids spending money on fencing more than can be avoided, and reserves his capital for other directions where it is more needed. We have known farms where large quantities of birds were maintained, the houses being placed out in the fields without any mixing of their inmates. If it is essential to keep the birds in confinement, and portable houses are desired, the form shown in Fig. 2 is a most valuable one, as the inmates are protected against their enemies. Fig. 3, fitted with a run in front, is equally useful, and is very easily removed. Cottagers can either use an existing outhouse, or build a lean-to, shown in one of the illustrations.

Adapting Existing Buildings for poultry is frequently the cheapest way of securing a good roosting-place, and when some disused outhouse, stable, coach-house, or shed is available this will save expense. The principles which must determine the adaptation of all buildings are that they be sufficiently large for the purpose, that they be dry and comfortable, and that they are not charged with impurities. Some buildings are not at all fit for this purpose, in that the walls or the floor are damp, which cannot be remedied, and when that is the case they should not be used. If the building is of brick or stone the walls should be well limewashed, so as to fill up all the interstices, which so frequently are harbourages for vermin. The roof should be made perfectly air and water tight, and the floor raised above the ground outside, so as to

Fig. 1.—Houses and Scratching-sheds, College Poultry Farm, Theale.

Built by Mr. W. Calway, Sharpness.

secure freedom from damp. If of wood, the outside should be well tarred or covered with felt, and the inside treated as already mentioned. Ventilation must also be provided for, with outlets for the fowls, also nests and perches, and the poultry-house should be well lighted. Any place will not do for the birds, as is sometimes thought; but these adapted buildings often make very good houses for poultry.

Aspect for the Poultry-house.—This is an important consideration, for the temperature will largely depend

FIG. 2.—PORTABLE HOUSE AND RUN.
(*Mr. W. Calway, Sharpness.*)

upon the position in which the house is placed. During the winter season it is desirable that the birds secure all the warmth they can get from the sun, and to do this their dwelling must be so placed that the early morning rays will fall upon the window. Thus, as a matter of general rule, it may be said that the house should face south or south-east; for under these circumstances it will obtain what sunshine there is during the winter season. There is, however, a modification to be recommended where the strongest winds are from the south-west, as is the case in the southern counties of England and Ireland, more especially the south-western counties, where Atlantic breezes blow very powerfully. There, in exposed positions, it is necessary to alter the aspect so as to face the east. Facing in this way the benefits of the morning sun are obtained and are of the greatest value. For a poultry-house to face the north in any place is exposing it to cold winds, and the inmates suffer accordingly. However good the doors and windows may be draughts will enter, and we know that the north

FIG. 3.—SMALL PORTABLE HOUSE.
(Mr. W. Greenwood, Bedford.)

is always the coldest side of the house. It must be explained that, where portable houses are employed, it is better to remove them where they will be shaded by trees in hot weather.

Chicken-houses.—Where chickens are intended to be reared on anything like a large scale, more especially if early in the season, a large roomy shed as a chicken-house should be erected. By such an arrangement the birds can be reared much more easily, and the cost is soon repaid. The best shape is long and narrow, with roof sloping to the back, as shown in Fig. 5. The front should be provided with large windows, so that the inmates can obtain the benefit

FIG. 4.—BARNARD, BISHOP AND BARNARD'S LEAN-TO POULTRY-HOUSE.

of whatever sun there may be, and they can be allowed out during the day, if favourable. In here the coops or brooders are placed, and the floor should be covered several inches with dry sand or gravel.

Fig. 5.—Winter Brooder-house, College Poultry Farm, Theale.
Built by the Portable Building and Construction Co., Manchester.

Cleanliness is of supreme importance in poultry-keeping, and from want of attention to this point a vast amount of disease is engendered. We were once asked to see some fowls which the owner said would not thrive, do what he would, and feed them as he liked. The explanation soon came, for the fowl-house floor was covered several inches thick with the droppings, trodden down but never taken away. Such a mass would at night, when the house was closed, give off an effluvia enough to poison anything, and we have found instances where poultry-houses were only cleaned out twice every year. The droppings should be regularly removed, the nests frequently renewed, the walls limewashed out two or three times a year, and swept down at least once a month. Food-vessels and water-fountains should also be daily attended to, and in fact everything done possible to keep the air sweet. Breathing a poison-laden atmosphere all night is most certain to induce disease.

Doors in Poultry Houses and Runs should be made sufficiently wide to permit a barrow or hamper to pass through. Unless this is done, there will often be a great waste of time, proving a constant source of annoyance. We remember once to have been in a large poultry-yard where this precaution had not been taken, with the result that everything had to be carried. It does not matter so much in houses as in runs, but is better in both. For the latter, we like to see the gates twenty-seven inches wide, as this permits an ordinary-sized barrow passing through comfortably. Gates and doors should be well hung, and made to fit properly.

Dry Shelters for Poultry are most important at all seasons of the year, but more especially during wet and stormy weather. Fowls appear to have a strange aversion to sheltering in their roosting-place during the day, and if they are not provided with a separate shelter will often prefer to stay out in the rain or cold. It is of no use fighting against this feeling, but it is better to provide for it. The dry retreat may not only be utilized for sheltering the fowls, but can contain the dust-bath. It need not be very large, but should be equal to the area of the house itself—in fact, may with advantage be half as long again. It should have communication with the roosting-place itself, so that the birds can pass from one to the other without going into the open air. A further advantage of such a shelter is that the birds can be kept under cover during wet or snowy weather. The front of a shelter-shed should be enclosed with wire netting.

Dust-bath.—It will be conceded by all who have had the least experience of poultry that a good dust-bath must be provided; for by this means fowls cleanse themselves. In the summer season, when at liberty, they can generally find a place by the roadside or under a hedge in which to disport themselves, but if in confinement or the season is wet and cold they cannot do so, and in any case it is desirable to provide a good dust-bath, which can be placed in the shed or within a special house or shed built for that purpose. The deeper it is, the more easily will the birds be able to use it, and we like to see a dust-bath fifteen to twenty inches deep. Nothing is better to put in it than fine ashes, which can be saved from the fire-grate, but fine, dry earth or wood ashes will do equally well, only they must be perfectly dry. If damp they are of no use. These will have a deodorizing effect, and are highly beneficial to fowls. There should be some disinfecting powder or black sulphur mixed with the ashes or earth to kill the insects. If, combined with a good dust-bath, there is care in the house in the matter of perches, and nests such as we recommend, insects can be kept down very easily, and all the trouble will be amply repaid by the way in which the birds will thrive.

Floors.—One of the most important matters in connection with a poultry-house is that the floor shall be perfectly dry. Unless this is the case the inmates will always be roosting in a damp-laden atmosphere—a fruitful source of disease. Bad as are damp or leaky roofs, they are not nearly so injurious as a floor so affected. And, again, if the floor is foul, air within the house can never be sweet. Floors should always be a little higher than the ground outside, so that the moisture will drain from, and not into, the house. They should be porous, not hard-caked, and an excellent plan for non-removable houses is to raise the house on one or two courses of brick, and fill in the space with broken brick, burnt ballast, coarse gravel, or stones, well beaten down. Above this should be placed three or four inches of sand, or, best of all, peat-moss litter, which is one of the most useful things introduced to poultry-keepers of late years. It is an excellent deodorizer, is soft to the feet, serves a long time, and makes a splendid manure. Where wooden floors are used, as must be the case when the houses are portable, these should be covered with three or four inches of peat-moss litter, both for the sake of the manure, and also that the house will not be so cold as would otherwise be the case. Uncovered wood, brick, cement, and stones are all very bad for poultry-houses.

Insect Pests.—It is the experience of all poultry-keepers that perches and nests are most prolific harbourers of insects and vermin, and many readers would be astounded if they realized that myriads of insects often infest their fowls when it is thought that they are entirely free from them. It is a commonly-accepted belief that where there are fowls there must be fleas; but this we beg leave to doubt. Certain it is that parasites can be enormously reduced in number, to the great comfort of the birds and their better thriving. But it must be acknowledged that there is a great difference in yards, so far as the number of these pests is concerned, for some fowls appear to be always infested with them, whilst others do not show many signs of their presence—at any rate, during the daytime. But if the owner would go at night into his fowl-house he would probably see a sight that would astonish him not a little; for there would be armies of these pests, alarmed by the light, returning with all haste to the crevices and interstices which form their lairs, and where they lie dormant during the daytime. Such a state of things is more often found in wooden than brick and stone houses, though if the walls of the latter are broken or uneven they may be just as bad; but as a rule they do not offer such facilities for the vermin. In a stone or brick building the walls should be primed, and in a wooden one the holes and crevices can be filled up with an admixture of soft soap and clay, to which a few drops of carbolic acid should be added, and this, when hardened, will do much to keep down the insects, which cannot thrive without a place of resort. Above all, no matter what kind of wall it may be, there should be a good thick coating of strong whitewash, well charged with carbolic acid, and it is always a good plan to have the walls as smooth as possible. Some glue or lump tallow mixed with wash will secure this result, if evenly laid on.

Nests, if neglected, are a great source of trouble in the way of insects. We have sometimes seen it stated that hens are led to lay away simply because their nests are infested with insects. Whether this is so or not we cannot say; but it is a very probable explanation, if the perversity of some fowls, like that of many people, needs any explanation at all. As a rule, nests are very much neglected, and boxes for this purpose are often made in the most unsuitable manner that could be devised. Close boxes (to clean out which would be a remarkable feat), the straw unchanged perhaps for weeks, and below all a mass of filth and dirt, is not an unfair description of nests we have seen, and it cannot be wondered that they form a harbourage for vermin. All such arrangements as sulphur-balls are in vain, useful though they may be, unless the nest itself is kept in good order. A capital form of nest, and one of the simplest that could be devised, consists simply of four pieces of two-inch scantling, so nailed as to form a perfect square of about fifteen inches. There is no top or bottom, it is just a frame of the size named, and no trouble is involved in the operation. Covered every now and then with a good coat of limewash, it would not provide any retreat for insects; but any box with the lid and bottom removed will do just as well. The most necessary thing is to remove the nest entirely once a week, to thoroughly sweep the ground below, and to supply fresh straw. Over the latter, when the nest is made, a little Izal may be sprinkled, if that is thought desirable; but, with the precautions already enumerated, it is scarcely necessary, though of course it cannot do any harm. It is better to place the nests on the dark side of the house, and raise them a few inches above the ground. Fixed nests are an abomination, and should never be employed.

Perches.—The best rest we know of is a concrete block let into the wall. We once saw this system capitally applied. The perch was of a **T** shape, and the three ends rested upon blocks of the material just named. The perch was easily removed, and as the only unevenness on each block was a slot in which the perch rested, there were very few facilities for vermin. This was in a brick-walled house, but in a wooden one the same plan could easily be adopted. For the latter, there is nothing better than a couple of strong iron eyes or staples fastened to each end of the perch, with corresponding hooks or catches on the be seen or got at. Perches are best made of fir-wood, about three inches in diameter, which, sawn in two, makes a couple of good perches. Failing this, we prefer a plain piece of deal about two inches wide and three inches thick, with all the edges rounded off, for if this is not done crooked breasts will ensue, and these are most objectionable. Perches should be all on the same level, and not more than two feet above the ground. It is a good plan to paint the ends of the perches with, or dip them into, paraffin once or twice a month, as this is generally disliked by the insects, and they will flee away from its smell. Should wooden rests be used for the perches, they may be treated in the same manner, and the comfort of the birds will be greatly enhanced, which will in its turn be greatly to the advantage of the owner, and thus the benefit be mutual.

Poultry-runs.—The question is often asked, How many fowls can be kept on, say, a plot of ground twenty yards square? The reason for this is not difficult to find, but to give an answer is not so simple. Much depends upon the ground, the kind of fowl to be kept, and how it is laid out. Heavy fowls bare the ground much more rapidly than do others, though there are some varieties which are much more active than others, combining an activity of body with a heavy frame. Plymouth Rocks, Houdans, and birds of that stamp need, therefore, more room than do many other varieties. If it is wished to keep all or a portion of the run in grass, it will not be possible to have more than six to eight birds on twenty yards square, and in this case it would be much better to divide the run into two, allowing them the use of each half on alternate weeks. But on runs of this size it is preferable not to have the whole of it in grass. The better plan is to lay down half in gravel or sand, and permit the fowls to go into the grass portion for about a couple of hours each day. In this way twelve or fifteen fowls can be easily kept on the space of ground named. It will be necessary to renew the gravel or sand once a year, but that is easily done, and there will be no fear of the ground becoming foul and tainted. Fig. 1 shows houses and runs arranged on this system. It is from this cause that so much disease arises, and soil unrenewed and constantly trodden over is certain to become tainted. Runs should be made long and narrow rather than square.

Ventilation.—By ventilation we mean a current of fresh air constantly passing through the house, but directed properly in its course. Draughty buildings are worse than no houses at all, and it would be wonderful if fowls kept in them did not become diseased. The walls, windows, door, and roof of every house should be air and water tight, and the ventilation so arranged that the fowls will never roost in a direct current of air. The simplest manner of securing ventilation is to put louvre-boards at each end of the gable, if there be one, and of course above the line of perches, so that when the foul air rises, as it will do, it can pass away at once. A few holes in the ends will answer the same purpose, and if some boards can be placed along the roof below the openings, so as to make an air-chamber, that will be an advantage.

Warmth.—Artificial heat should never be employed in poultry-houses. If they can be placed in a sheltered position, or adjoining a dwelling-house, stable, or cowshed, it will be sufficient. Wooden houses can be made more comfortable if thoroughly tarred outside, fitting a layer of brown paper between the second and third coats. Felting is also good for this purpose.

CHAPTER V

HINTS ON BREEDING POULTRY

Influence of the Parents.—It is an accepted fact in connection with all kinds of breeding stock that the male parent influences external qualities and the female internal. Thus, if we desire handsome birds, or birds of a certain type and shape, we must look to the father, but if we want good layers, reliable mothers, or rapid fatteners, we must depend most of all upon the influence of the mother. In many instances too little attention has been paid to the value of the mother in breeding, for whilst we do not for a moment doubt the value of the father, yet he has only his own position, and cannot possibly exercise influence both for himself and his mate. In both sexes we should look externally for a well-balanced development, and the more certain in type the better. But in the case of any defect this should be compensated by an excess of that quality in the opposite sex. Hens for breeding should always be of good shape, compact, large in comb, and, as far as possible, from mothers that are known to be good in the quality sought for. A good layer will generally produce good layers, and other qualities are transmitted in the same way. The following directions as to the selection of a cock bird are culled from an American paper: 'Do not select a male bird with stilted legs, or with hock-joints crowding backwards and inwards, or that lack firmness; neither one with legs near to each other, or crooked, or with slim thighs, or with narrow breast and back, for such a bird lacks muscle, and has not sufficient room for vital organs; nor one with slow motions or of a cowardly nature, soft, loose-feathered, with scant tail plumage or meagre hackle; nor one too large, fat, and clumsy, devoid of vim and vigour.

Eggs from hens running with these coarse, indifferent males hatch with uncertainty, and chicks of such a sire are not as fine and serviceable as those resulting from the use of lively, amorous cocks of any variety.'

Age of Breeding Stock.—It is inadvisable as a rule to breed from fowls under twelve months old, and to secure the healthiest chickens one parent should not be less than two years old. A fowl cannot be said to have stopped growing until the first moult is passed, which generally takes place when it is seventeen or eighteen months old, and consequently until that natural change is completed it is practically in a state of immaturity. Much enfeeblement of stock and deterioration in point of size has arisen from the evil practice of using immature birds in the breeding-pen. When chickens are to be bred for killing, and for that purpose alone, cockerels and pullets may be used as breeders in order to secure fertile eggs early in the season, but in that case no harm is done, as the progeny is not to be used for further reproduction. Some varieties of poultry can be used in the breeding-pen for several years, whilst others, especially of heavy kinds, become sterile much earlier. First-rate birds that have proved their value as breeders should be used as long as they are of any service in this respect, but as a rule it is better not to keep stock after the third year.

In-and-In Breeding is a vexed question among all classes of breeders, and requires the most careful consideration. There can be no question that close breeding has had to be adopted in order to obtain purity of type in all branches of live stock, and that without it many of our present races would have had no existence. But so far as the *industrial poultry-keeper* is concerned, there is very little need for this practice. There is a sufficiency of pure-bred races for his use, and whilst it is most desirable to maintain purity in the stock, excessive attention to externals is

White Leghorns.

neither necessary nor desirable. The dangers of close breeding are great, for any weakness will be accentuated. One great reason for deterioration of fowls in the United Kingdon on farms has been the way in which interbreeding has been permitted.

Mating for Breeding.—Unless the breeding stock is properly mated, it is impossible to expect anything like uniformity or to hope for success, whatever the object of the breeder may be. There is a very erroneous impression amongst many poultry-keepers that only fanciers need take any trouble in the selection and mating of their breeding stock. Certain it is that to the care exercised by fanciers is due much of the success attained by them. Without such care it would have been impossible for them to have secured anything like the quality and high character of many varieties of the domestic fowl. And this fact applies to all races of domestic animals. Where the object is to produce economic qualities the same laws apply, for it is only by attention to the tendencies and qualities of the breeding stock that laying can be improved or that the quality and quantity of flesh carried on the frame can be developed. In mating, therefore, the first thing is to have a distinct end in view, whether this be the improvement of colour, shape, feather, or of the more profitable qualities, such as laying and table properties.

Whilst it is desirable to this end that animals be selected with something like uniform tendencies—for instance, if layers are wanted, to select both cocks and hens from families in which this quality runs—at the same time it is to be remembered that absolute uniformity is never to be found in Nature, and if there are any minor deficiencies in one or other of the fowls, they should be compensated in the opposite sex, as already suggested. This is, however, more requisite in breeding for external than internal points; still, it is a very potent factor, and one which should be regarded by every breeder. We may exemplify what is here meant by saying that supposing layers are being bred, and that there is a desire to increase the size of the eggs laid, in that case the stock-birds should be selected from varieties or families noted for producing large eggs. It is in this manner, namely, by studying the special characteristics of fowls, and utilizing these characteristics to secure the desired result, that success in this and all similar pursuits is attained.

Number of Hens to a Male Bird.—The number of hens to be allotted to each cock depends greatly upon several considerations, and no certain rule can be laid down which will apply to all varieties and all seasons. In the first place, the period of the year must largely determine the number of hens running with a cock bird, for whilst one male will fertilize the eggs of ten to twenty hens in mild, open weather, it is probable that six or eight hens will be quite enough early in the season, when it is cold and stormy. Therefore a commencement should always be made with the smaller number, as others can be added so soon as doing so is thought desirable. On the other hand, the number of hens must not be too small, for infertile eggs as often result from the running of too few as of too many hens with the male bird. It is necessary to carefully watch this point, for under no case should the hens be distressed by the attentions of their mate, which can easily be known by their backs becoming bare of feathers. If such a result is seen to be the case, more hens should be added to the pen. There is also a great difference in the virility of the varieties of poultry. It is not at all uncommon for young cocks of the more active breeds to fertilize the eggs of ten or more hens, but in the heavier varieties six are generally quite enough as a maximum. There is, however, a vast difference in birds of the same breed, and each poultry-keeper must determine this question largely for himself, being guided by the appearance of his hens, and whether the eggs laid are fertile or not. If infertile, it will be better to take one or two of the hens away, and that will generally prove whether too large a number has been the cause. Frequently there is another reason for infertility of eggs, to which reference is made in the next paragraph.

Treatment of the Male Bird.—Often the infertility of eggs is due to a cause little suspected, namely, that the cock does not obtain sufficient food to eat. This may be surprising, but it is none the less a fact. Cocks are nothing if not gallant, and they seldom or ever attempt to feed themselves until their harem is fully satisfied. If the supply of food be a limited one, and for various reasons it is most undesirable to allow food to stand about, there will be nothing left for him, and, as a consequence, he does not receive the support which is necessary. It is a rule with many poultry-breeders to always feed the male bird by himself after the hens have been satisfied, and this is a plan which can be recommended in order to prevent infertile eggs and the consequent disappointment through them. The poultry-keeper can easily find out whether the cock is obtaining what food he requires, but it must be pointed out that there is less danger of this sort of thing when fowls have free range than in confinement.

Healthy Stock.—The chicken-raiser has three great difficulties to face. First, in too many instances, want of stamina in the birds themselves. This is a most serious question, and one that cannot be dealt with after the chickens are hatched. It is a fact, the proof of which can be seen around us every day, that the more delicate animals are frequently the most prolific breeders. But it is equally true that their progeny are often born only to die, or, what is worse, to linger on in weakness and misery. Therefore, and we urge this to the full extent of our power, it is far better to kill off birds of a weakly nature than by coddling and nursing rear them to adulthood, and perhaps allow them to engraft their weakness in an enhanced degree upon another generation. The first consideration on the part of the poultry-breeder who wishes to succeed with his stock is that they be healthy.

Breed only from the Best.—A very general mistake made, especially by farmers, is breeding from the entire stock. By such a plan it is impossible to secure that improvement which ought to be the aim of every poultry-keeper. Even where pure male birds are introduced every year, and of the right class, by mating them with the entire flock of hens there is no control over the hens, and it is often the case that the worst of the entire stock are reproduced because they chance to lay when eggs are wanted for sitting. The breeder should select the best of his hens—best in accordance with his requirements—and, dividing them into flocks of ten or a dozen, mating each flock with a male bird, place them in separate houses. This is an economical method, as it obviates the keeping of more males than are absolutely required.

CHAPTER VI

SITTING HENS AND THE HATCHING OF CHICKENS

Addled Eggs.—This is a term used to denote those eggs which have been fertilized, but from one cause or the other have not come to maturity, the embryo dying during the process of incubation. There is a considerable amount of misconception as to addled eggs. An egg which has never been impregnated with a living germ does not become addled, and the fact of its being so addled is a proof that it has been fertilized.

A barren egg will simply be fusty when sat upon, retaining its colour; but a fertile egg, in which the embryo dies during the first few days of its development, becomes rotten. This rottenness is, therefore, as a rule a proof of fertility.

Artificial Incubation.—Incubators are most valuable where operations are conducted on anything like a large scale, and soon repay their cost. By their

Fig. 6.—Hearson's Incubator.

means chickens can be hatched at almost any season of the year, and certainly whenever fertile eggs are obtainable; and now that these machines have been brought to so great a state of perfection, they can be relied upon to do their work well and with a minimum of trouble. It takes no more labour to attend an incubator holding fifty eggs than a couple of sitting hens, and sometimes less, whereas the result is, or should be, twice as great. Hens' and ducks' eggs should never be put into the same machine, as the latter hatch better at a lower temperature than the former. For hens' eggs the heat should be maintained from 103½ to 104 degrees, whereas for ducks' eggs 102 degrees will secure better results. Incubators should be placed where they are not subject to great variations of temperature.

Broody Hens are more or less to be met with at all periods of the year. Some breeds have lost the sitting

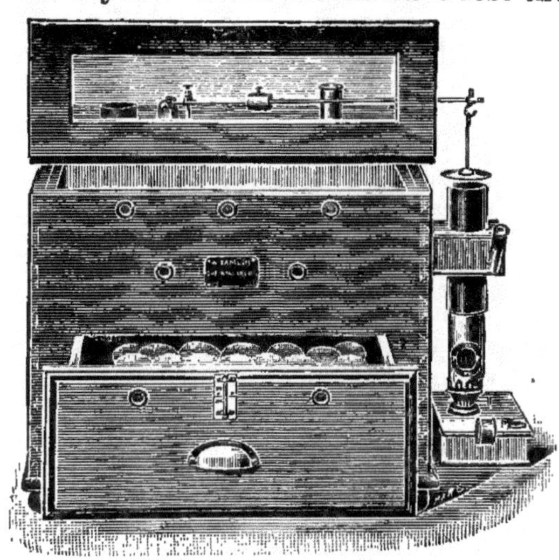

FIG. 7.—TAMLIN INCUBATOR.

propensity, it having been bred out, but in others it is very strongly developed. The signs which betoken a desire to sit are a constant occupation of the nest, and that peculiar, but indescribable, sound which has gained in some districts the name of 'clucking.' Amongst the regular sitting varieties it is usual to allow each to bring out two or three hatches in the year, and with birds that it is intended to breed from the next season it is a desirable plan to allow any hen who commences 'clucking' a nest in the late summer, for this hastens the moult, and brings them into fit condition for laying through the rest from egg production thus given. Sometimes it is thought necessary to break a hen off who desires to sit, as she is not needed for that purpose, but we do not approve of the system. In this case the old cruel methods of dipping in water, or starving, need and should not be resorted to. It is usually sufficient to place the hen in a coop with a barred floor, upon which she cannot form a nest, if even she had the materials wherewith to do it, and keep her there for two or three days in sight of the other fowls, which, however, she is unable to join. In all varieties—save, perhaps, Cochins—this is usually the most effective way.

Chilled Eggs.—In the early hatching season very frequently the breeder is much troubled by sitting hens abandoning their nests. Our experience has been that this more frequently takes place during the prevalence of east winds, whose occult influence upon breeding operations has never yet been defined. But a sudden spell of cold will bring about the same result. When one out of several hens does this sort of thing, her eggs can be distributed amongst the other hens, and the matter is thus easily arranged; but if six out of eight or nine do so, other tactics have to be employed. It is at a time like this that the value of an incubator is experienced, for the deserted eggs can be put into and hatched out in the machine, or kept there until other hens are obtained. Failing this, the eggs should first be taken into the house, and placed for a few minutes in water heated up to 105 degrees. This will have the effect of reviving the chilled embryo, or chicken. If there is not another hen or an incubator at hand, we should recommend the placing of a hot-water bottle, or a couple of water-bottles, in a roomy box with the eggs around, but the bottle must first be encased in flannel, or the heat will be too great. We

Black Minorcas.

have several times kept eggs two or three days in this manner, and in one instance some four or five out of a dozen eggs so treated hatched out. Of course, the water in the bottles must be renewed from time to time, in order to keep up the temperature, which should be maintained as near 104 degrees as possible. It is wonderful what an amount of cold eggs will stand, and unless the shells were actually cracked with frost, we should always make an attempt to save them in one or other of the ways described. We have known eggs left by the hen for thirty-six hours in severe weather when so treated hatch out all right, but, of course, the period of incubation was greatly extended, and they were a few days late in hatching.

Cooling the Eggs.—It is commonly accepted as a fact that all eggs, whether in an incubator or under hens, should be allowed to cool once or twice a day, and in practice it is found that eggs so cooled hatch stronger chicks than if this point be neglected. Eggs can be hatched in a machine without cooling if the regular temperature is lower than 104 degrees, but there is much truth in the saying that 'a close sitter is a bad sitter.' Hens should be compelled to come off for feeding once a day, lifting them by the wings if needed, and eggs in an incubator should be cooled by opening the drawer from ten to twenty minutes, according to the season of year, twice a day. If the air is very cold, a piece of flannel should be placed over them. With a hen, the nest should be examined during this time of cooling, and if she has fouled her nest, or an egg is broken, the nest should be remade, and all the eggs washed in water heated to 105 degrees, the hen's breast and thighs also to be cleaned.

Food for Sitting Hens should always be hard corn, never soft food, as the latter is apt to induce looseness of the bowels, and so cause fouling of the nest. An excellent plan is to have feeding-cages, in front of which food and water should be placed, so that the hen can reach them easily. The ground should be covered with fine ashes to form a dust-bath, and thus keep down insects among her feathers.

Hatching-boxes.—The best hatching-box we know is 15 inches square and about 18 inches high, but without bottom. On one side, which we shall call the front, nail a strip of wood 6 inches wide across the bottom, and to this hinge a door, covering the remaining portion of that side, and fix it by a catch. If a few inch holes are made in the sides near to the top, then the box is complete; and when in use the hen is secure from molestation, and cannot leave her nest except the door

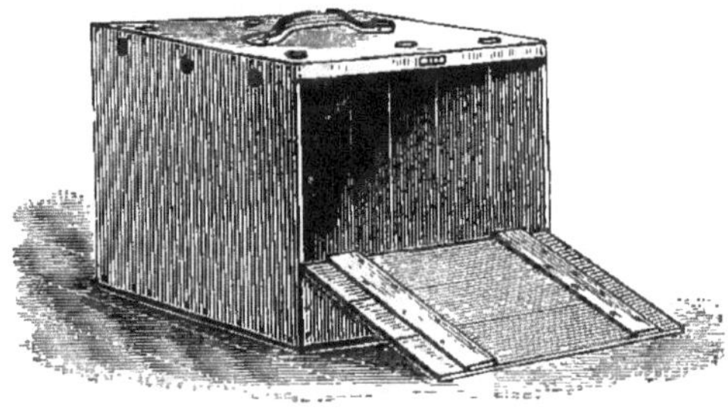

FIG. 8.—HATCHING-BOX.

is opened, the door itself being a ledge for her to use in going in and coming out, or the box may be made with a lid on top, in which case the hens must always be lifted off. In France baskets are often used instead of boxes. In the first place, it will be necessary to cover the floor of the hatching-room with fine ashes or earth a foot thick, and upon this the boxes are placed, all facing towards the centre.

Making the Nest.—To make a nest, put enough fine earth into the hatching-box to fill up within an inch of the top of the ledge; then slightly hollow out the soil, and lay upon it some soft bedding, either oat or beaten wheat straw. Care must be taken to see that the corners are filled up with earth, so as to prevent the eggs rolling away from below the hen, and thus being chilled.

Moisture is essential to successful hatching. If the atmosphere be damp, and the layer of earth below the hatching-boxes is kept moist, sufficient will be obtained in this manner. Duck and geese eggs may, but hens' eggs should never, be damped with a piece of flannel or sponge. The better plan in dry weather or a dry place is to pour hot water *around* the outside of the hatching-box, so as to keep the earth damp, and the hen's body-heat will draw it up to the eggs. In incubators, attention to proper and regular supply of moisture is essential.

Moving Broody Hens.—Many bear removal quietly, and we have known sitters sent fifty miles by train and then take to the new nest of eggs as readily as if they had never seen any other. But in order to do this a hen must be carefully packed. Our plan has been to buy a hamper—a new-potato basket is best, if it has a lid; make a nice nest in the bottom, put two or three chalk or spoiled eggs in it, and then place the hen gently in, when she is ready for her journey. But there are hens also that will not bear removal, and often refuse to take a nest, even when this is but a short distance from their own abode. Such as these require very careful treatment, or the broody fever will go off, when the hen must be returned to the laying-house.

Places for Hatching.—If only one or two hens are to be set, they can be accommodated in any empty shed or outhouse; but where several will be so engaged at one time, it is better to devote a room to that purpose. The less subject to variations of temperature, the better will be such a place, and this applies also where incubators are employed. An empty stable makes a good hatching-room. Nor need it be perfectly dry, so long as the atmosphere is pure, moisture being an important element in hatching. A dry, hot place is most unsuitable of all. Hens sit better in a dark room, and should never be where they can see or hear their companions. They should be disturbed as little as possible.

Selecting a Sitting Hen.—Sometimes when eggs are plentiful broody hens are just the reverse. A note of warning is therefore necessary, as it is undesirable to take the first birds offered without regard to their suitabilities, and whether they can be trusted or not. Of course, when broody hens are very scarce, then some risk may be run; but at the same time it must be remembered that eggs are valuable, and the giving of these to an uncertain sitter may result in their entire loss. We have often found it better to wait for a few days to secure a trusty mother than commit the care of eggs to one who, though broody at the time, is not of a breed as good for maternal duties. It often happens that broody hens which have to be sought for are mongrels and of no definable breed, but very little practice will enable anyone to tell whether a hen is likely to prove a good mother or not. In the first place, all feather-legged birds may be depended upon in this way; but if heavily feathered and clumsy in build, there is another danger—namely, that the eggs or chickens may be crushed through the awkwardness of the mother. Cochins and Brahmas are sometimes very exasperating for this reason, and thus we do not care to give eggs (pullets' eggs, at any rate) to pure-bred birds of either of these breeds. It must be remembered that the majority of eggs laid in the spring are from pullets, and in these the shells are generally much thinner and weaker than those laid by hens. But half-bred birds—say Brahma-Dorkings or Cochin-Dorkings—can be used with safety, and to the first-named we must give the credit of being the best sitters and mothers of all. They are large, and can cover a goodly number of eggs, are careful but not clumsy, will fight for their broods, but yet allow an attendant to handle them, and altogether can be thoroughly recommended. Some poultry-breeders keep a number of these birds for the purpose of hatching and rearing, and they are very wise in so doing, as there is great danger in having to trust untried hens. Plymouth Rocks, Langshans, Wyandottes, and Orpingtons make excellent sitters and mothers.

Settling a Hen.—When the hen has been secured and the nest made, a few spoiled or chalk eggs should first be placed in it. At night put down a wire pen in front of the box, with food and water, and into this pen the hen. Here she may be left an hour or two, with a lantern throwing the light, so that she can see her food, the nest, and very little else. As a general rule, it will be found, if the door of the hatching-box is left open for the purpose, that she will go on to the nest and very soon be comfortably ensconced thereon. If she is wild, it will be better to remove the lantern before the door of the box is shut, else she may be frightened and fly out. She should not be interfered with or let out of the box for twenty-four or thirty hours, when, probably, she will be all right, and more difficult to tempt off her nest than otherwise. If she is settled properly, the right eggs may be substituted for the others, though in any case it may be as well to wait another day until the uncertainty has passed into certainty. It is of no use being impatient in a matter like this, as impatience may result in the loss of a valuable sitting of eggs.

Testing Eggs.—There are several kinds of egg-testers made, most of which are on the same principle, though there is considerable variation in the form. The object of the tester is to darken the surroundings of the egg, and permit light only to pass through the egg itself. In this way the clearness or opacity of an egg can be determined very easily after it has been subjected to the incubation process for a few days. Many persons can test eggs by holding them in the hand, thus dispensing with any tester. To do so, the egg is held in the left hand by the finger and thumb, the other fingers being closed to intercept the light; the right hand is laid against the now-exposed right side of the egg, and in this simple fashion a most effective tester is made; or a hole rather smaller than an egg can be cut in a piece of cardboard, and the egg laid against this hole—of course, between the eye and the light. In a darkened room the sunlight will be enough to test eggs; but usually it is found more satisfactory to do this at night, or in a darkened room, using a candle or lamp. Recently, a lamp has been introduced by Messrs. Dawson and Co., of Birmingham, by which eggs can be tested very early in the process of incubation. This test proves whether eggs are fertile or not, barren eggs being clear, whilst fertile eggs show after the second day. But the sixth day is a convenient time for testing, as the egg shows opacity in the centre.

CHAPTER VII

CHICKEN-REARING

Care versus Coddling.—A MOST important consideration is how to rear the chickens without injuring them by kindness of treatment. There can be no question that the best way to secure hardy stock is to let a hen set herself when and where she likes, allow her to bring up her brood in her own way, and give her all the liberty which she wishes. But poultry-keeping upon such a system is only possible to the few. Were these conditions to be insisted upon, three-fourths of the present poultry-keepers would have to abandon the pursuit. More than that, the results of such a system would be unsatisfactory, both as to numbers produced and to profit. There is a great deal of nonsense written respecting natural methods of keeping and rearing poultry. Domestication alters very greatly the habits and requirements of the fowl. Such is the lesson taught us by all masters of natural history. As well might we say that the horse should be treated in exactly the same manner after generations of service to man, when he has adapted himself to the altered conditions of life, as if he was wild upon the plains of Arabia, as to claim for fowls this so-called 'natural treatment.' We do not mean by this that they should be pampered and coddled, but simply that we may recognise the change brought about, and deal with the birds in the light of our experience.

Having said this much, we are anxious to warn very strongly against the opposite course. It is not too much to say that 90 per cent. of the diseases from which our fowls suffer are due to the foolish methods of feeding and management. A celebrated surgeon once said that 'man can live in any climate, can perform almost any kind of labour, can enjoy perfect health, and, bar hereditary disease or accident, need not die until he is eighty years of age, if he will regulate his diet according to the demands upon his system, to the conditions of his life, and observe sanitary laws.' And the same eminent surgeon said, 'Nine out of every ten of my patients would never need my assistance did they but watch what they eat.' These great truths apply equally and with as great force to poultry-rearing. What we are anxious to urge is that from the very first the chickens shall be treated in a common-sense manner, that they be not coddled, and that they be fed on nutritious yet plain food. This does not apply to birds destined to an early death, but to those which it is intended to rear to adulthood.

Brooders or Artificial Mothers.—The artificial rearing of chickens proved successful long before artificial hatching, but the introduction of incubators has certainly made it more popular. There are, however, many people who hatch in machines and rear under hens; others who hatch under hens and use arti-

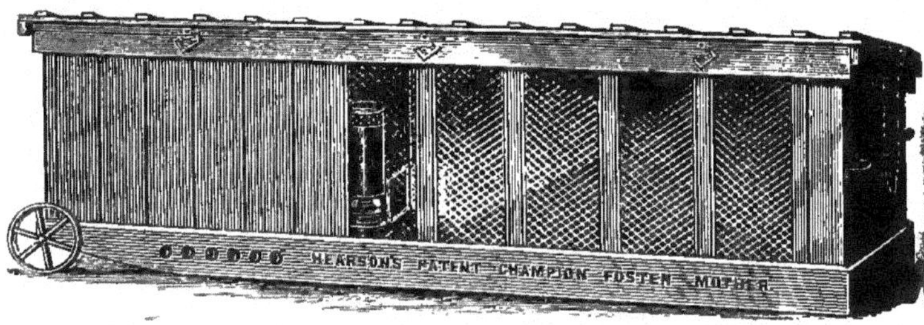

FIG. 9.—HEARSON BROODER.

ficial mothers; and further, those who both hatch and rear by natural methods, but use mothers when the hens desert their broods. All these plans may be adopted according to the pleasure or convenience of the breeder. We used an artificial mother long before

FIG. 10.—WESTMERIA BROODER.

an incubator—in the first place, because we found it simple to manage with the limited space we then had at disposal, there being no fear of fighting between hens, and we also found that the chicks actually did better than when going about in the usual way. Those who rear under hens, even if they hatch in a

LANGSHANS.

machine, do so because they have plenty of space for the broods, and also under the impression that the hens are better able to look after the chicks and obtain for them worms and things of that kind. Of the larger forms, the Hearson and Westmeria are types. They are really heated chicken-houses, each

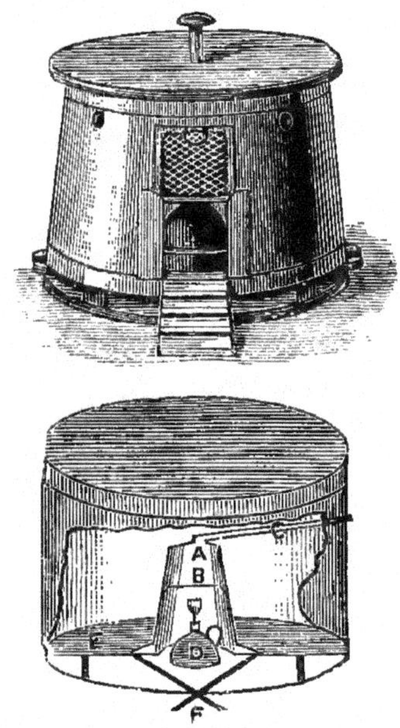

Section showing Heating Arrangement.
A.—Heating cone. B.—Disc to throw heat to side of cone.
C.—Air-pipe. D.—Lamp. E.—Platform. F.—Regulator.
FIG. 11.—CALWAY'S PATENT CHICKEN-REARER.

provided with a warm sleeping chamber and a covered run. They are fitted with wheels, and can be easily moved about, and chickens can be reared at any season of the year. The Calway rearer is an excellent machine for smaller numbers, which we have found reliable. It is both simple and inexpensive. With all these it is most important that the chicks shall not be overcrowded and the mother is not too hot, or it will make the chicks tender. This is a mistake often made, and we must warn our readers very strongly against it. If the mother be heated to 80 degrees at first, this is hot enough in cold weather, and 10 degrees below that will do in warm weather.

Chicken-coops.—When chickens are reared under the ordinary conditions they should be cooped out in the open, except very early in the season, when this would be too much exposure for them. The system of cooping itself is so well known that it does not need either defence or description; but as much depends on the coop itself, it is desirable to say what are the best kinds. Amongst the oldest forms is that which is usually called the Sussex coop (Fig. 13), which at one time was almost universally employed. This is triangular in shape, and presents at the front the appearance of a large A. It generally stands about 24 inches or 27 inches in height at the apex, and is the same width at the ground. The sides and back are solid, but the front is made of upright laths, one of which is loose, so that, when taken out, the hen is free, the laths being sufficiently apart to allow the chickens to pass between. In fine weather and sheltered positions this coop is all that can be desired, but otherwise it does not allow of a good shelter-flap at the front. For this reason we prefer the square-shaped coops, as with them there is not this fault to be found, and a drop-shutter at the front will be a shelter both against sun and storm, besides being convenient for entirely closing in the coop at night as a measure of protection to the inmates. The top may be either gabled or have a simple slope from front to back, but should in either case project well over the sides, so as to carry the rain right away beyond the coop. Whether the coop shall have a wooden floor or not depends upon the place whereon it will stand. If the soil is dry, it is much better to dispense with the floor except very early in the season; but where the soil is damp or cold, such an addition to the coop is indispensable. In all cases where used, it should be made separate to facilitate cleaning, and fit within the sides of the coop so as to keep out the wet, being raised on two pieces of three-inch scantling, to which it should be nailed. If this scantling projects six or eight inches in front of the coop and the floor also, an admirable feeding-board will thus be provided. Another form of coop (Fig. 14) can be recommended, namely, a double coop, with a partition dividing it exactly in half, and bars separating the two. One half is closed in all round, having a solid door in front, whilst the

FIG. 12.—SPRATT'S PATENT REGISTERED SHELTER COOP.

other half has a lath front for the passage in and out of the chickens. The whole of the front of the former half forms a door for cleaning purposes. For early broods this is certainly the best coop we have seen.

Chicken-houses.—The value of a house for chickens has already been referred to (Chapter IV.), more especially for early hatches. Only in this way can very early broods be reared, unless brooders such as have been mentioned are employed. A chicken-house can be utilized after the breeding season for fattening purposes, or as an abode for a lot of laying hens after they have been separated from the male birds.

Cleanliness is of supreme importance in chicken-rearing. Coops, brooders, and dishes should be regularly cleaned, and the two former should be frequently limewashed out, certainly between every brood. Rotting food, stale excrement, and dirt are fertile sources of disease.

Conditions for Chicken-rearing.—Amongst the most delightful and interesting parts of work in the poultry-yard is that of chicken-rearing, but perhaps there is nothing less understood. The loss of young life in all animal species must ever be great, but the accentuation of modern civilization has but conducted into another channel the law of 'the survival of the fittest.' There is, however, this difference: that whereas under

FIG. 13.—CALWAY'S SUSSEX COOP.

natural conditions it is the physical weakling who succumbs, under our artificial state of existence it is rather the digestibly enfeebled who fall victims. And as doctors tell us that thousands of children die annually from over and ill feeding—far more, in fact, than those slaughtered by neglect—so in the case of chickens the mortality is enormous from this cause alone. Below we deal with this subject, as it requires careful attention, more especially from those who are compelled to keep their birds in confinement. Consequently it must not be imagined that all the troubles of the chicken-breeder are over when birds are out of the shell, for in many instances they are only just beginning. We have often said that anyone can hatch chickens, by which is meant that, given ordinarily fair conditions, it is not difficult to secure good hatches; but the rearing of the birds needs constant care and foresight to avoid troubles which are so often met with. Of course, if we have healthy stock, a suitable place for rearing the birds, and no enemies to fight against, all goes well. As a rule, one must be content with what one can get in the way of accommodation with limited space, and therefore it is imperative that the strictest attention be paid to the rearing of the birds in order to prevent disaster.

FIG. 14.—DOUBLE COOP.

That chickens can be successfully reared under unfavourable conditions is proved without doubt, but only when constant regard is paid to details.

Cooping Hens with Chickens.—When there is sufficient space, there can be no question that it is a much better plan to coop chickens out in the open except very early in the season. By so doing we most nearly adopt the natural method. Except in very severe weather, the coop may be permitted to stand exposed on all sides, though the shelter of a hedge or a hill on the lee side will be a great protection. For the first week or two a hen should not be permitted to wander about with the little ones, as she is very apt to overtax their strength by leading them away too far; but when they have thus far grown, and there is sufficient space at command, it will be better to give her liberty during the day. Where the amount of space is limited, and there are several coops near together, the hens must at all times be cooped, or they would fight and vent their spite on the unfortunate youngsters of other broods than their own. One most essential matter in rearing by means of coops is that these small abodes shall be removed to fresh ground every day, or that the fresh ground shall be brought to them, for it is most fatal to the chicks if this is not done. By a judicious disposal of the space at command, if it be limited, all danger from fouling of the ground can be avoided, and where it is impossible to keep any green turf by reason of the numbers reared on it, the use of sand or fine earth will prevent mischief. We do not lay the same stress upon grass as do some authorities, and if it came to be a question between an open ploughed field and a grass run, we should prefer the former. They will get as much green food as they need even in such a field, and the other kinds of food obtainable from the broken earth will bear down the scale immensely in its favour. A hen should be left with the chicks seven or eight weeks, when she will discard them.

Feeding Chickens.—Whilst feeding is at any period of life a most important matter, there can be no question that it is of special importance during the earlier stages. Therefore, if it is desired that the birds shall grow healthy and strong and develop in the directions desired, they must be fed accordingly. No animal which has been neglected during this period of its growth can ever attain either the size or the stamina of those more favoured. It is therefore essential that the feeding be well attended to, and that the food be right in kind and quality. Poultry-breeders should study the qualities of different foods, and feed accordingly. This subject is dealt with fully in 'Poultry-Keeping as an Industry for Farmers and Cottagers.' In feeding chickens it is needful at the outset to remember that they must be fed in conformity to the object in view. For instance, if they are intended to be killed at an early age, they should be fed in a different manner than if they are to be reared as layers or breeders. The food suitable in the one case would be fatal to success in the other. Food which would go to produce bone and muscle would be unsuitable for table chickens, and *vice versâ*. During the first few days the best food we know is eggs boiled hard, chopped fine, and mixed with an equal bulk of breadcrumbs, the whole being moistened with milk, but not sloppy. This may be given entirely for three or four days, when there may be alternated, and then substituted, a variation as follows: Spratts' Patent Chicken Meal, poultry rice, and oatmeal. The Spratts meal and oatmeal must be properly steeped and the rice well cooked. These must be continued if it is intended to rear the birds to adulthood; but if for killing, there may be alternated with these for the first two months boiled rice, Indian meal, in which meat or suet may be mixed, and wheat or buckwheat. For the first two or three weeks birds will be unable to eat whole grain, and dari or cracked buckwheat is the best to be given, but after that time they will be able to take all the smaller-sized grains. Chickens should be fed every two hours the first week, every three hours for the next month, and afterwards four times a day until they are fully grown. They should never be fed to repletion, but always given just as much as they will eat readily. Food should not be left standing, as it soon sours, and sour food is fatal to chickens. Sweet milk may be given to young birds, but not too freely, as it is a rich food.

Plymouth Rocks.

First Days of a Chicken's Life.—In all forms of nature treatment during the earliest stages of existence has great influence upon the future. If an animal, fowl, or plant be neglected, ill-fed, or exposed to conditions for which it has not been prepared by parental influences, it will either succumb, be dwarfed, or so enfeebled that its life is not what it might otherwise have been. This is emphatically so in the case of chickens, many of whom have to die an early death or to live an emasculated life because of their early treatment. Very often this takes the form of too great kindness, and it is incalculable how many thousands of chickens die every year 'killed by kindness.' There are several general principles which may here be laid down for the guidance of chicken-rearers, and which will prevent any of the troubles named if they are observed. In the first place, Nature has provided the food for the first twenty-four hours of a bird's life by means of the yolk-bag, which is absorbed just before batching. Therefore, to force the chick to eat until it is ready for food is almost certain to induce digestive troubles. So soon as they are ready they will eat, which is usually at the end of the first day. Much must depend upon the place where chickens are reared, for if there is abundance of run they will find many elements needed by them. In confined places bone-meal, fine gravel and coarse sand, flint, grit, lime, etc., should be supplied freely, and as much green food as possible.

Shelter for Chickens.—In rearing there is a happy mean, by which the young birds can be dealt with in a common-sense manner. They can be helped forward, made to meet the needs of their owner as to time and place, and yet, provided that they are bred from healthy stock, will not be coddled. We firmly believe in exposing the chicks at the outset to the weather, not unduly, but sensibly. An exception must be made in the case of very early birds bred for marketing in the spring, but for all stock hatched during the regular hatching season it is better to place them out in the open from the first. It is true that under this plan very weakly ones will die off, but their room is better than their company, and it is very seldom that there is any satisfaction in rearing them. At the same time, it is equally true that some of the birds who would have pined and died if coddled will be stimulated into a more stalwart life by this treatment. Shelter they will need against wind and rain in winter and spring, and sun in summer, and for this purpose a hillside, a thick-set hedge, or a wall, are all most useful, and in case of severe storms it is very desirable to have a good shed into which the coops and their occupants can be moved. Of course, we only mean to say that the chickens shall be subjected to the ordinary variations of our climate. A shower of rain, even a little snow, can do no harm if the chicks have a warm and dry shelter and are fed on suitable food.

The Older Chickens.—Too often poultry-rearers, when they are having a succession of broods, are apt to leave those about six to eight weeks old pretty much to themselves, under the mistaken impression that the little youngsters need the more care. This is a great mistake, for just when the hen is deserting her brood they need special attention. Their food must be good, they must be protected against the weather, and also prevented utilizing their newly-found freedom to their own hurt. Whether reared under a hen or in a brooder, a cold mother is very useful at this period. Bonemeal mixed in their food and plenty of grit are essentials, and they should have as much grain, such as buckwheat or crushed oats, as they like, giving warm meal only once a day. When they have been left by their mother for two or three weeks, they may be placed in a roomy house, and allowed to sleep on a good bed of peat-moss litter or straw.

CHAPTER VIII

FEEDING AND FATTENING

Food Values.—The importance of food is unquestionable. It is the fuel which keeps the machine at work. Therefore it must be constantly supplied, or the machine will collapse. It is not, however, with the necessity for supplying food we have to deal. So much is universally admitted. The trouble is that, instead of food not being given, it is usually supplied too liberally, or of a rich and stimulating nature. Infinitely more harm is done by giving too much than too little, and for one case of the latter we find a score of the former.

What the fowl requires is, first, a food that shall supply the daily expenditure of force which takes place in every living creature. This expenditure varies in fowls as it does in human beings. As the man whose daily toil is hard and physically laborious requires food greater in heat-forming qualities than does he whose labour is light, and as the man who lives in a cold climate needs food that will supply the rapid exhaustion of heat in his body, which form of exhaustion the resident in warmer spheres does not feel to the same extent, so it is with fowls. And the food which may be given in one place will be dangerous in another. Not only so, but birds at liberty, by the very amount of exercise they obtain, can assimilate and eliminate much richer foods than fowls in confinement. But even granting all this, there can be no question that the plainer the food the better, unless some special object is in view, such as fattening for the table, when it is not fed long enough to cause harm. There is an abundance of choice amongst the various grains at our disposal, so that there need not be a level uniformity in feeding. Bread is the staff of life, but we like to have a change, yet return to bread with greater zest for the change. Variety should be obtained as far as possible in the feeding of poultry. This is not so important with fowls at liberty as with those whose freedom is restricted. The former find a great variety for themselves, and the grain supplied is but part of their food. But the latter are unable to do this, and what they need has to be artificially supplied.

Barley is usually regarded, and rightly so, as a valuable food for poultry. It is freely given, both whole and ground, and in either way is to be recommended, as it has most of the elements which are desirable in feeding. Nor is the cost any hindrance to its use, as in most parts of the country good barley can be purchased at a reasonable price. The best samples are used for malting, and are consequently more expensive; but the ordinary samples are worth using or buying, as there is full value for the money in the results. From analysis it appears that barley is very rich in fat-formers, but in this respect it is not nearly so much so as wheat, and the larger degree of woody fibre and mineral matter has the effect of counterbalancing the apparent excess of fattening matter. Whole barley may be given either alone or mixed with other grain. Barleymeal is, of course, the cereal ground to powder, and is better if mixed with, say, one-half to two-thirds of good pollard. In buying barleymeal it is important to see that the meal is pure, as there is a great deal of mixing with it other and cheaper substances, it specially lending itself to adulteration. For this reason it is desirable whenever possible to buy the barley and have it ground, rather than take what is offered in shops.

Bonemeal, or Ground Bones, is a most valuable assistant to the chicken-breeder, especially those who have not the advantage of very extensive grass runs. Where the breeding is conducted in more restricted places it ought certainly to form part of the diet of all the growing chicks. Bonemeal should not be used for chickens which are to be marketed or killed at an early age. In their case the object is to secure as much flesh with as little bone as possible, and early maturity should be sought for. But with chicks that are destined to a longer life either as layers or breeders, or even as capons, it is essential that there be a large frame. The best form is when fresh bones are ground up, and these are very valuable for egg production. The Furness Bone-machine is useful for preparing this food, as by its use fresh bones can be cut up into small pieces suitable for all classes of fowls. In America, these are largely employed for poultry, especially laying hens.

FIG. 15.—FURNESS' BONE MACHINE.

Buckwheat is a very useful food for poultry-feeding, and in France and Belgium especially it is very largely used. It is a native of Russia, but is now grown in almost all other parts of the Continent and of America. Buckwheat is stimulating in its nature, and is therefore most valuable for feeding laying hens and young chickens. It is largely used in France for fowls when they are being fattened, but is then mixed with fat and milk, which, for the purpose named, impart the elements it lacks. It gives a good colour to the flesh, and is in every way one of the most suitable foods for poultry.

Condiments for poultry have been condemned wholesale by some writers, but this is as foolish as it is to advocate their regular use, regardless of time and season. There can, however, be no doubt that the use of condiments is, at certain periods of the year, most beneficial, supplying elements which are then much needed. But in warm, dry weather they should not be used under any circumstances, as they are not merely useless for the purpose to be achieved, but

Fig. 16.—Parish's Cooker.

they do positive harm to the fowls. The admixture of a good condiment, such as 'aromatic condiment' and others largely sold, with the soft food, early in the season for chickens, and during cold or wet weather for adult fowls, gives an internal heat which is a preventive of disease, and frequently the means of securing eggs when the hens would not be laying. But condiments need to be most carefully used, and ought never to be employed when there is not actual need.

Cooking Food.—A most economical method of preparing food is by cooking either meal or whole grain, which when in this state are very digestible. For this purpose cookers should be employed, such as the Parish or Lancashire, an oil-lamp or coke-stove, according to the size, supplying the heat. By cooking many materials, such as clover hay, potatoes, or

FIG. 17.—HEARSON'S CRAMMER.

vegetables, can be used as poultry food, which would not otherwise be the case.

Cramming.—Both in France and England there is much done in the direction of cramming. Various methods are adopted, but the principle is the same, namely, that fowls are forced to consume a given quantity of food per diem, the object of which is to add to the quantity of flesh they carry ere being killed. In France the system may be said to have been perfected, as it is carried out on an extensive scale. The process of cramming is performed by hand, by a funnel, or by machine. In the first named the food is made up into finger pieces, and after being dipped in milk is forced down the throat of the fowl until the crop is full. For funnel-feeding the food is made liquid. When a machine is used the mouth of the bird to be fed is placed over a nozzle provided for the purpose, and either by pressure of the foot or a turn of a handle sufficient of the food is forced into it. Full details as to this system are given in 'Poultry-Keeping as an Industry for Farmers and Cottagers.'

Damaged Grain is generally sold for poultry, and its great recommendation to many purchasers is its comparative cheapness, or, rather, low price. We do not say that damaged grain may not be good for poultry, but it is not necessarily so, and considering the amount of nutrition, which is the thing we have to consider, this class of feeding is often the most expensive. Much sold is quite unfit for food, and should not be given under any circumstances whatever, as it is really only fit for manure. If the grain must be used, it should first be boiled, and the scum rising should be removed. In this way it can be partially improved. Sometimes grain can be bought which has been slightly damaged with water, or has been badly harvested, which is little the worse for feeding; but the grain sold in ports has been heated in the ships, and is not so suitable. Tail wheat does not come into the category of damaged grain, and it is, as a rule, even better for fowls than full-bodied grain, especially for laying stock.

Dari is a grain which has come into favour for feeding, more especially chickens. It is a small white seed from the plant called Indian millet or Guinea corn, which is largely cultivated in India, China, Africa, Italy, the West Indies, etc. It appears to be specially valuable during the breeding season, and is most useful for young chickens. The nutritive power is high, and dari is a grain which can be recommended.

Fattening.—By the term 'fattening' we do not mean the mere laying on of an excessive quantity of fat or oil, but rather the addition of just as much fat as is necessary, and increasing the quantity of flesh. Fattening softens the flesh and ripens it, and herein its great value consists. In order to attain success in fattening, two things are absolutely essential: first, that the food-supplies shall be suitable; and, second, that the conditions under which the fowls are kept shall help towards this end. The ordinary method of feeding is insufficient, for if fowls are allowed in the open air there will be a great loss through atmospheric changes and exercise. Fattening is an artificial state, and it is as necessary to the ripening of fowls as a hot-house for grapes in this country. With regard to food, the best kinds are ground oats, barley, buckwheat, rice, and maize. Ground oats are chiefly employed in this country, and are the finest food we have for the purpose. In France many of the fatteners use buckwheat-meal, which, with the addition of a little suet and milk, makes a splendid food for this purpose. In fact, nearly all meals need fat added to them. Food should always be given in the form of meal, as it is so much easier to digest, and the food should be given milk-warm. Regularity in feeding is most important, and on no consideration should any left be permitted to stand on to the next time of feeding. Twice a day ought food to be given, and as much allowed as they will eat readily. The period lasts usually from fourteen to twenty-one days. It is better to keep birds in separate cages, just large enough for them to stand in, and in a dark

WHITE WYANDOTTES.

room, which should be kept as even in temperature as possible. They must be kept scrupulously clean.

Green Food.—In one form or another green food is essential to fowls. The best kind of green food is undoubtedly grass, for this contains elements which are most of all needed. But there are places and seasons when this cannot be obtained, and it is then important to find a substitute. All kinds of green crops, vegetables, and most of the roots can be so employed with advantage. The best of all is lettuce, especially for young chickens, and some of the most successful breeders are those who give their chicks as many spring lettuces as they can consume. Next we should place young cabbage, and afterwards all kinds of vegetables. Turnips may be given whole and raw, allowing the birds to peck at them; but harder vegetables, such as potatoes, should be boiled and mixed with the soft food. In the winter season, when grass is poor and vegetables are scarce, a bundle of hay for the birds to peck at is a good substitute. It is best if suspended just within reach of the fowls, and so suspended affords amusement and employment to those in confinement.

Grit.—The first part of mastication takes place in the crop, where the food is softened. Much disease results from the want of proper materials, for if food is not ground in the gizzard, it either passes through the stomach without its elements being absorbed, or makes an undue tax upon the system, either of which is most injurious. The materials required—which fowls at liberty obtain for themselves—must be sharp and hard, so as to be capable of grinding the food. For this purpose fine gravel, broken shells, road-scrapings, and very coarse sand are all useful. Of late broken flint has come into regular use, and it is splendid for the purpose when broken to a proper size. It is sold by many poultry-food purveyors, and the quantity which fowls will consume shows how they appreciate it. A box of grit should be available for fowls of all ages from babyhood.

How to Feed.—It must be remembered that it is quite possible to provide the right kind of food, and yet by giving too much of it arrive at the same undesirable end as if rich food were supplied. Some persons contend that the right way is to keep a constant supply of food before the fowls, but with this opinion we cannot agree. The best method of feeding is to supply food twice or thrice a day, in troughs, and the removal or ceasing to give more food when the fowls appear satisfied. Feeding to repletion is always bad, and doctors tell us we should rise from the table with an appetite, *i.e.*, not actually hungry, but feeling that we could eat more if we wished to. We should endeavour to apply the same rule to our fowls, and the doing so will be an effectual preventive of overfeeding, with its subsequent evils. The giving of tit-bits every now and again during the day is to be deprecated, as is the striving to tempt the appetite by dainty dishes. As a rule, we have found that, when the appetite of a fowl fails, the best remedy is the withholding of food altogether. It may be offered, but if not accepted should be taken away again. In a day or two Nature will right herself, and then plain food will be most acceptable. But it is better to obviate all these troubles by forethought.

Indian Corn or Maize is one of the commonest poultry foods, because of its cheapness and apparent value, but it is a grain which should be most carefully used. Maize is very rich in fat-forming materials, and, as a consequence, fowls that have not sufficient exercise to eliminate the fat from the system, if fed on maize, become internally coated with fat, which, deposited on the various organs, induces disease in one form or the other. Liver disease, apoplexy, and some forms of heart affections, are the result of feeding on maize. When fowls are at liberty, and are of a variety that will wander freely, or are kept in a very exposed, cold situation, they may have small quantities of maize in cold weather; but for poultry in confinement, or of the heavier varieties, it ought not to be used at all.

Natural Food.—Animal food is, to a large extent, the natural food of fowls, and when at liberty they consume a considerable amount. The commonest forms are worms, insects, and other things of a like nature. The quantity of worms which a few fowls will consume is very surprising, and to witness the way in which they will follow the plough shows how much they appreciate this kind of food. There can be no doubt that fowls fed on such a diet are far richer and higher flavoured than those fed on grain alone. This is one reason why the fowls of France are so splendid in flavour, for they are permitted, in the wine districts, to wander at will amongst the vines, and in that way obtain the food they desire most of all. When poultry are kept in confinement, it would be injurious if they were permitted to have as much animal food, nor do they need it. To eliminate from the system the elements derived from these rich foods, it is essential that the fowls be more exposed, and have an abundance of exercise. This they do not obtain in confinement. It is desirable, therefore, to be most careful in giving animal food artificially. A proportion is desirable by reason of the large amount of nitrogenous elements contained in meat, but it is only a small proportion, except in winter, when there is much greater demand upon the heat reserves than at other seasons of the year. Household scraps are the best of all, but if this sort of thing has to be purchased, the best we know of is the Crissel made and sold by Spratt's Patent, and similar meat meals. By giving too much animal food laying can often be forced for a time, but will ultimately result in disease. Still, in the winter season it is very valuable indeed.

Oats.—Of all grains, there is no doubt that oats are

the very finest poultry-food, in that the requisite elements are found therein in the best balanced form. In some parts of the country sound, plump oats are not always obtainable, but where they can be secured, and food has to be purchased, they should be preferred to any other, more especially for growing chickens and laying hens. In Surrey and Sussex ground oats are largely employed for fattening. In these the oats are ground up entirely, husks and all, very sharp mill-stones being employed. It is the perfection of poultry-food. Oatmeal is not so good, having been kiln-dried, but may be used for chickens.

Preparing Soft Food.—Soft food should always be given crumbly moist, never sloppy. Porridge may be good for human beings, but it is bad for fowls. To mix meal, place it in a pail or whatever vessel will hold it, making a hole in the centre, into which pour *boiling* water. Then stir with stick or wooden spoon, and add more water as required, until the whole is a crumbly mass. Feed in dishes, or make into balls, which throw down to the fowls one or two at a time, ceasing to do so when they are satisfied. When food is cooked it must be dried to a crumbly mass by meal.

Water is a most important part of the diet. A running stream solves the difficulty at once, but only very few poultry-yards are so favoured as to have this at command. If fountains are used, they must be kept perfectly clean, and to secure their being so no fountain should be used into which the hand cannot be put. Disease is often caused by impure water—disease of the most subtle and deadly character. Water should be often renewed in fountains, but if in open troughs, these should not be exposed to the sun. Very often the water-supply for fowls is regarded as a minor consideration, but it has much more to do with their health than is generally supposed.

Wheat.—For laying fowls wheat, especially of the thinner sorts, is an excellent food if given whole; but when ground is too floury, and needs the admixture of some coarser material, or it becomes pasty. Tail or small wheat is by far the best for poultry, and if sound is almost equal to the best samples of oats.

The question of fattening, killing, dressing, and marketing is dealt with exhaustively in 'Poultry-Keeping as an Industry for Farmers and Cottagers.'

CHAPTER IX

DUCK, GEESE AND TURKEY BREEDING

Ducks.

The most profitable time for selling ducklings is in the early spring. There is also a large demand for ducks in the summer and autumn, and especially in districts contiguous to watering-places and tourist resorts. This latter demand can only be provided for by breeding ducklings in May and June. It is to be remembered that ducklings mature very rapidly, and if fed properly are fit to be killed when eight weeks old; in fact, if not killed before they are nine weeks old, they must be kept for some time longer, as at the age named they begin to shoot their quill feathers, and when that commences they must be kept until the process is completed. As this means feeding for six or eight weeks without any corresponding recompense in the way of increased weight, it will easily be seen that doing this will take away much of the profit, if not actually turn it into a loss.

Rearing Ducklings.—Ducks are very hardy, and ducklings remarkably easy to rear under ordinary conditions, but it must be confessed that in many places ducks cannot be reared during hot weather.

We have sometimes known in the months of June and July scores of ducklings lost, from three days to a fortnight old, and without any apparent cause. They hatched out well, seemed to thrive for a few days, and then died off without any premonitions of disease. It was not for a considerable time that this state of affairs was altered, nor until we had almost begun to think it was of no use breeding ducklings during the summer. Afterwards we found that the place and housing were to blame. Ducklings in the early part of the year will thrive almost anywhere, but in the summer they need plenty of shelter and running water. The best shelter is that formed by trees and bushes, and a cool, shady plantation or wood is admirable for the purpose. Without this we should not advise anyone to go in for duck-rearing in warm weather, but with it and a nicely sheltered stream there would be considerable profit to be made at that time of the year. Many a place, such as we have described, which would be too damp and cold even in March, can be utilized in this way, and we should strongly urge those who have such at their command to give it a trial.

Housing.—At any season of the year ducks need much less protection in the way of housing than do fowls, and a mistake is often made by amateurs in keeping ducks within close sheds or houses. In summer an open-fronted house will be sufficient for old birds, and a very rough-and-ready one will do for ducklings. If there is no fear of night marauders, the latter may be accommodated beneath an open shed, boards being placed to the height of a foot above the ground. This, with plenty of clean straw, will be quite sufficient. If, however, night marauders are feared, a roomy shed should be employed, made with either a thatch or a wooden roof. The sides may be solid, in which case free ventilation should be given, or, what is even more to be preferred, the sides may be solid to a foot above the ground, and above wooden laths or wire netting. Either of these will afford ventilation, and that is what is most needed. Overcrowding in small houses must be avoided if success is to be attained.

Feeding.—A different system of feeding must be adopted from that which is suitable in the early part of the year. The food should be as plain as possible until within a fortnight of the time of killing. Spratt's meal mixed with barleymeal in equal parts for the fortnight, and then barleymeal three parts and Spratt's one part, alternated with boiled rice, the rest of the time, will be found the best diet. Too much even of this should not be given, for the ducklings will, in such a place as we have named, find a great deal of their own food. For the three weeks previous to killing, the ducklings—that is, from five or six weeks old to killing—should be kept in an enclosure attached to their house and fed on boiled poultry-rice, with a little greaves or meat-scraps mixed therein. This will cause them very rapidly to put on flesh, and they will be found in splendid condition for killing in the time named. If the yard is also covered with straw, the droppings will saturate into it, and thus form a splendid manure. Of the varieties recommended for summer breeding we should place Aylesburys first, and a cross between this breed and Pekins next. For rather later requirements the Rouens should be chosen. Further particulars are given as to these breeds in Chapter III.

Geese.

In this country, geese during certain seasons of the year meet with a large demand. It is true that turkeys have taken the first place, which was once occupied by geese; but commensurate with this change there has been a great increase in population and

Aylesbury Ducks.

Rouen Ducks.

in wealth, which has compensated for the change. Green geese are regarded as a great delicacy, and at the Christmas season vast quantities of fattened birds are sold. The finest of these are home-grown, principally produced in the Eastern counties of England, in Cumberland, and very fine ones are reared in Ireland. We have a large supply from the continent of Europe, and of late years large shipments have come from Canada; consequently, the prices for inferior qualities, such as we receive from abroad, have been very low, through the flooding of the markets, and geese can generally be bought at 6d. or 7d. per pound. These lack the flavour of our home-grown geese, and are evidently not very carefully fattened. The lesson to be learnt is that, in order to succeed, the very best should be produced, as these realize the best prices.

Geese in Norfolk.—If we look to the great geese-raising district of England, we shall learn something of the method adopted in the place where the best are bred. In Norfolk the trade resolves itself into two great branches—the green geese and the Michaelmas, the Christmas trade being really a continuation of the latter. Some farmers do not attempt fattening, which process is left to the dealers, who lay themselves out for it. In March and April the dealers begin to obtain their gosling supply from farmers or cottagers near the commons, and, as a rule, these goslings are from five to six weeks old and very thin. They are fed for six or seven weeks under stages on barley-meal, maize, wheat tailings, and brewers' grains mixed, which food soon makes them ready for the green-goose market. These stages or pens give them sufficient room to move about, but not too much, and they are kept there for all the period, being allowed out now and again to have a bath and for cleaning their pens. Michaelmas geese take their places under the stages in August, and the endeavour is made to have them ready for the market before the Irish and the Dutch supplies begin to arrive. On turnips geese are found to be capital substitutes for sheep, and when a fattener has a turnip-field, he not unfrequently hurdles a portion of it and eats it off with them. They first eat the tops, and then the bulbs of the softer turnips, and when they are put upon swedes, the man in attendance gives each turnip a chop. Geese will eat a field cleaner than sheep, and their manure is equally good. When fed in this way, they need nothing more than a trough of water, and the finishing process consists in putting them under stages for a month, and feeding on steeped oats or barley-meal, brewers' grains, and potatoes.

Rearing.—It is well known that the management and food during the first few weeks of any animal's existence determine very largely its future size and the rapidity of its growth. Of course, there are great variations in breeds as to their capacity for putting on flesh freely and rapidly, but very much can be and is determined by early treatment and management. No bird which is neglected and half fed during the earlier stages of its existence can possibly thrive, and it is essential to begin from the first with proper feeding. Goslings are about the easiest of all domestic poultry to rear, and when once hatched require very little looking after, after the first few days. When hatched, as they may be under the mother bird or hens—whichever is most convenient—put them in a roomy coop or crate, and it is better not to give them a large run at first. The coop must be situated in a sheltered position, for the sun's rays are very fatal to goslings. Of course, if there are plenty of bushes or trees about the matter is very simple, as the coop can be placed under the lee of these; but if the situation is open and bare, some shelter must be improvised. The coop can then be made with an eaved roof, but in addition large hurdles, in which has been interlaced furze, should be provided, and if freely scattered about they will be welcomed by the goslings. It is an excellent plan to cover the top of the coop with furze, as that keeps the coop cool. In all cases the coop should be bottomless, for goslings are unable to hold their feet on a wooden floor, and are very apt to injure themselves by slipping about. It is not necessary to go to much expense in the making of coops, for they can be built at a reasonable cost, as they are not required very strong. Goslings are not usually hatched until the weather is open and mild, and are not delicate by nature. When about ten days old they can be liberated from the run, and will prove themselves splendid foragers.

Feeding.—The early feeding must all be in the direction of building up the frame on which the flesh may be laid. The best food from the first is wheat and oatmeal, or barleymeal and wheat. The whole grain should be scalded and dried up with meal. Variations may be made by giving ground oats, and by boiling potatoes and drying them up with meal When about two months old they may be fed on raw grain and sliced potatoes, and this will suffice until the time of fattening arrives. They are also very fond of young grass, young green onions, chickweed; and an early cabbage hung within their reach in the run will be highly prized; in fact, green food is the most important part of a gosling's food. No other special provision is necessary, except the giving of water, which must on no account be forgotten. The vessel should be of such a nature that the birds cannot get into it, as they will assuredly do if it is open to them. Therefore, one of the closed fountains is the best, if there is not a running stream. For autumn killing the Embden is to be preferred, being quicker in growth than the Toulouse, which is most suitable for the Christmas trade.

Turkeys.

Of all the varieties of domestic poultry, turkeys have the character of being the most difficult to rear. They need careful attention during the earlier stages of their growth, and if this attention be not paid the probabilities are that there will be a very serious loss. In the first place, it is important that birds raised should have sufficient space, as it is absolutely useless attempting to rear turkeys in a limited run. That is where many poultry-breeders make a mistake, and when failure meets them they put it down to every cause but the right one. It must be remembered that the turkey is really the last of our domestic fowls that has been brought from its wild state under domestication, and that the recent infusion of fresh blood from America introduces again some of the wilder instincts, so that, as the American wild turkey is allowed to roam where it will, confinement causes the birds to pine and die. In turkey-rearing, what is most of all to be desired is that the soil shall be porous, carrying off the water speedily, and not be heavy clay. Warmth is also essential, and a cold, bleak place would be unsuitable for this purpose. We believe that some districts are more suitable for rearing turkeys than are others, because from the nature of the soil insects and grubs are most plentiful, and as the turkeys feed on these to a large extent, they thus thrive better than where they are scarcer.

Rearing.—The strange fact about turkey-rearing is, that whilst adult birds are hardy and can be maintained without much trouble, the young birds are the most difficult of all poultry to rear. The reason for this we have never seen satisfactorily explained, and what we have suggested above is the nearest attempt we have seen made as to its solution. When the young turkeys have passed the first few weeks of their existence, or the time known as 'shooting the red,' they may be regarded as out of danger, but up to that time need the most constant and watchful care. If kept in coops—under ordinary circumstances this is the plan most usually adopted—it is essential that they be put on fresh ground every day. They are wide rangers, and their size makes them able to bare the ground much more quickly than chickens, so that the removal of the coop to fresh grass is of the utmost importance. We have seen it recommended that if a field is given up to the rearing of turkeys, part of it should be hurdled off and the grass kept short, so that when the other portion is bared this will be absolutely fresh and untainted. In this case, if the field were large enough for the number of birds to be reared thereon, by the time the second portion has been fully used the first would be again fit to put the coops on. Here we may mention that it is most essential that young turkeys should not be permitted to wander amongst long grass, for that would simply be fatal to them. They seem unable to withstand wet plumage, and the grass in the fields where they are being reared should be kept short. All this means trouble, but without trouble it is impossible to rear turkeys. It is consoling to know that this constant oversight only needs to be given for the first six or seven weeks of the young turkey's life, for it is at that age when the 'shooting of the red' takes place, after which they do not require more care than ordinary chickens. In rearing turkeys, however, it must be borne in mind that the reward is proportionate to the trouble, for the prices realized for these fowls are very good indeed. The Bronze American turkey is most popular by reason of its large size, but the Cambridge Bronze is rather better in quality of flesh.

Feeding.—An Eastern-counties breeder has written to say that food for the newly-hatched birds should be principally hard-boiled egg, lettuce, onions or nettles chopped up, and a few breadcrumbs; to this may be added rice, boiled in skim milk (if quite sweet), a little suet or greaves, and in cold or wet weather a sprinkling of cayenne pepper, with grain of all sorts as the birds get older. We have also seen churn milk recommended instead of the skim milk, and it ought to be good for this purpose. A fresh site for the coop every day is most important, as we have already stated. After the birds have passed the critical stage of their existence, they should be fed upon sound, substantial food, but rich diet is positively injurious. Barleymeal, barley, and small wheat should be the staple food, but with the barleymeal a little scrap-cake or greaves might be given. A very small quantity of maize may be used without any harm resulting therefrom, but this is a grain which the poultry are safer without. We note that an American writer attributes much of the non-success in turkey-breeding to the use of maize, and warns breeders against the use of this grain. It is altogether too rich and fattening, and should only be used as already directed. To feed off, shut up in a roomy shed for about six weeks, and give three times a day ground oats or barleymeal mixed with skim milk, and milk to drink. Sliced mangels, turnips and swedes, and cabbage, are useful, and there should be an abundant supply of sand, lime, ashes, and brickdust at hand. Cleanliness is most essential to success in turkey-rearing.

For further particulars as to ducks, geese and turkeys, see 'Poultry-Keeping as an Industry for Farmers and Cottagers.'

CHAPTER X

General Management and Diseases

General Management.

Disinfectants are of very great assistance to the poultry-keeper, but are too frequently used in a reckless and useless manner. A bad smell is not necessarily made innocuous because another more powerful odour is set up. Too often when, from want of proper attention to cleanliness, or in consequence of overcrowding, the nose is offended by bad smells, it is thought to be sufficient if some strong disinfectant be used. Disinfectants are good, but they have their proper place, and whilst they will do a great deal in the way of purifying, they must not be expected to work miracles. Therefore, when they are used, other steps should be taken to keep houses and runs sweet. We always have by us a good supply, and a slight sprinkling on the floor of the house, in the nests, or amongst the ashes which form the dust-bath, greatly assists in keeping down parasites. Fortunately, the best disinfectants, whilst deadly to insect life, are not harmful to animals.

Egg-eating.—The cause of egg-eating is sometimes difficult to determine, but usually it is simply an evil habit. Fill a blown egg with a strong mixture of mustard and cayenne pepper, closing up the ends with gummed paper. Leave this in the nest, and if the hen tries to eat it she will get a dose she does not like. Sometimes it is due to the want of shell-forming materials, the supply of which will stop the trouble. Nests should always be raised about a foot above the ground, as there is less temptation to egg-eating when so placed.

Feather-eating.—Some birds contract the habit of

feather-eating, due probably to their being kept in confinement. Sometimes the presence of insects in the feathers is an exciting cause, and a good dust-bath with carbolic powder mixed with fine ashes should be provided. At other times it is a depraved habit, due to unnatural conditions, and this is most difficult of cure. Give employment to the fowls by burying the grain in a heap of rubbish, and suspending a cabbage by a string just within reach. Fowls at liberty are seldom given to feather-eating.

Marketing Eggs.—The matters which chiefly need attention are speedy placing before the customer, and

FIG. 18.—POCOCK'S EGG-BOX.

the opening of fresh outlets. Hitherto, strange to say, the importance of speedy marketing has been scarcely realized, in spite of the fact that there is an almost unlimited demand, and at high prices, for perfectly fresh eggs, by which we mean those under three days old, and within reach of the producers. In some of our great centres of population it is scarcely possible to obtain really new-laid eggs. An egg has no more right to be called 'new-laid' at three days old than would milk deserve to be called 'fresh' at the same age. Until the importance of rapid marketing is realized no great improvement can be expected in our home trade, either as to prices or sales. Further, fresh outlets must be sought for. It is of no use running on old lines or in old ruts. Prior to the doing so, it is essential to have a regular supply, both winter and summer; but with a continuous supply there are many retailers, supply stores, co-operative societies, hotels, etc., who would be only too glad to have the opportunity of securing eggs that they can place with confidence before the most fastidious of their customers, and for which they will pay at excellent rates.

FIG. 19.—HONEYCOMB EGG-BOX.

Marketing Poultry.—It is essential that fowls reach the market where they are to be sold in as good a condition and as promptly as possible. So far as the prices realized, much will depend upon the time when fowls are marketed, and upon their size and quality. Anyone who has a good supply of poultry and takes care to market at the right season can command a good price; but here, again, if there are only a few birds for sale, and these are sent irregularly, it is scarcely to be wondered at that the returns are unsatisfactory. Much may be done in the way of opening up new markets. But it must be pointed out, and that very strongly, that there must be quality and regularity of supply on the part of producers. Fatted fowls, for reasons already stated, command the best prices.

Moulting.—The moulting of birds is an operation which usually takes place every year, and is the casting of one set of feathers and the replacing of these by entirely new ones. There is a constant change of and growth in the feathers all the year round, but only during the moulting is there anything like regularity in the process. The time during which birds are moulting is a somewhat critical period for all birds, its nature depending very much upon the stamina and constitution of the breed; but it will be readily seen that the strain upon the strength of the fowl, which the formation of new feathers must cause, cannot fail to be a heavy one, and during this period it is necessary to pay rather more attention than usual to them. Give warm food once a day, with a little aromatic compound in it, the grain to be buckwheat and barley, of which a handful can be given in the middle of the day. To the drinking water add a little sulphate of iron and sulphuric acid—say two pieces of the former the size of nuts and ten drops of the latter to a gallon of water, and, of course, this must be renewed every day, as fresh water is one of the most important of all things. A little boiled linseed mixed in the soft food is also of great assistance.

Poultry Manure.—The value of poultry droppings as a manure upon the land has been acknowledged for many years, but there is still much prejudice against it. When we consider that Peruvian guano is simply the deposits of wild birds, we fail to see any reason for this prejudice in the case of fowls. Analysis has proved poultry manure to contain many of the elements in which the best guano most excels. It can be guaranteed pure, which is not always the case with artificial

Toulouse Geese.

fertilizers; and even considering the amount of moisture it contains when in a fresh state, it is worth half, at least, what is usually paid for manures. The value of poultry manure has been proved over and over again, and farmers who keep fowls in portable houses can always tell where they were located the previous year by the heavier crops to be found round the place. In addition to the droppings that fall upon the fields, there is usually a large quantity to be obtained in the houses, which must be preserved until the proper season. This can be done in different ways, either by drying and keeping in barrels under cover, or by putting a thick layer of fine soil upon a hard bed of earth, and then layer by layer of soil and manure, always leaving the soil uppermost; when peat moss is used this will do equally well. By this means all the value of the manure is preserved, and any unpleasant odour is effectually prevented. For fruit and vegetable growing it is the best manure that can be obtained.

Diseases.

Apoplexy.—Many birds die suddenly, and without any apparent cause, generally being found dead under the roost. This is usually due to overfeeding or the use of Indian corn, which should be avoided. If one bird dies in this manner, reduce the food of others at once, and give a strong aperient.

Bronchitis.—Highly-bred stock, or delicate breeds in a cold, damp place, are sometimes affected with bronchitis, which can easily be known by its resemblance to that affection in the human subject. It is possible to effect a cure temporarily, but, as the complaint is very likely to return, it is better to get rid of such birds.

Bumble-foot may either be a soft swelling or a corn in the foot. If the former, let out whatever matter there may be, bathe the place with arnica lotion, wrap up in a soft cloth, and keep on straw until better. If a corn, cut out, touch the part with caustic, and treat as before. Bumble-foot may be due to high perches.

Canker.—This is very difficult to cure, and is an ulcerated state of the mouth, due generally to bad conditions. Wash the mouth three times a day with strong alum-water or borax and water. Mix sulphur or powdered vegetable charcoal in the soft food and feed carefully.

Cholera.—In some parts of the country large numbers of fowls have been swept off by an epidemic to which the name of fowl cholera has been given. It is most contagious and very deadly. Bad housing, want of ventilation, and foul water, are all exciting causes. There is practically no remedy, and all that can be done is to remove the cause. Birds affected should at once be buried in quicklime.

Cold.—The symptoms of cold are running at the nostrils, watering of the eyes, and sneezing. It is by no means difficult to cure if taken in time, but if neglected may speedily turn to roup or consumption, both of which in chicks are speedily fatal. When the secretion is seen at the nostrils, let the bird be at once separated from the others if old enough, and put in a warm, comfortable place; the food given must be good and nourishing; the medicine to be very small doses of tincture of aconite, and in addition a little camphor to be put in the drinking water. If, however, the chick is with the hen, the whole brood had better be given a warmer place.

Cramp.—Chickens of about six to eight weeks old are very subject to cramp, chiefly due to damp, as it is found mostly in birds reared on a clay soil. They first begin to show it in the feet; the toes curl up so that they cannot spread them out; then the whole leg is affected, and they walk upon their knuckles.

The first sign is a little lameness in walking, and as soon as this is noticed remove to a dry-boarded floor covered with straw, and keep there until better. Foment the legs well with hot water, and if the birds are large enough put strips of flannel as bandages round the legs and feet, or rub the legs with turpentine and put near a warm fire at night. The food must be good and stimulating.

Cropbound.—When a bird is seen to be cropbound, which is known by its crop being full and hard, pour some warm oil or milk and water into the crop, and then gently knead it for a time; but if that fails to remove the mass cut open the crop, making a perpendicular slit sufficient to get three fingers in. Remove all the contents of the crop, wash it out with warm milk and water, and oil the lower passage. Next sew up the slit with silk or horsehair, stitching the outer and inner skins separately. Feed on sopped bread for a few days.

Diarrhœa may be caused in various ways, such as a chill, want of cleanliness, improper feeding. It can be very largely prevented in young chickens by the use of bonemeal, but of course it is an absolute necessity to keep all food and water vessels sweet and pure, as nothing brings it on sooner than sourness in dishes. As soon as the droppings are noticed to be loose, some boiled rice should be given in which has been mixed a little powdered chalk. This will generally have the desired effect, but diarrhœa is the way in which Nature often gets rid of objectionable matter, and if too sudden a check is put upon it direct harm will be done. If, however, the diarrhœa increases, in spite of what we have already suggested, a drop of chlorodyne in half a teaspoonful of water must be given to each bird, or a drop of brandy can be used instead, if handier.

Diphtheria or Diphtheritic Roup.—A most deadly disease, and not dissimilar to chicken cholera, for which it is sometimes mistaken. It is very contagious, and may be communicated to man. When the growth across the throat is discerned, it will be the better plan to at once kill the bird, so that if possible the disease may not be spread.

Eggbound.—When a hen is eggbound, oil the vent and parts in order to soften them, and also inject a little of the oil. Probably this will be sufficient in the majority of cases, but a dose of castor oil will often assist. Should the egg still be retained, hold the vent over a jug of hot water in which a few drops of tincture of iodine have been mixed.

Gapes only attacks young birds, the gaping of the mouth being caused by the presence of small red worms which lodge themselves in the windpipe, and if not removed the chick soon dies of suffocation. Many measures may be adopted as preventives. Good food is necessary, food of a sound nutritious nature, and some poultry-breeders have been accustomed to anoint the heads of the chickens soon after they are dry with a preparation made as follows: Mercurial ointment, 1 oz.; pure lard, 1 oz.; flour of sulphur, ½ oz.; and crude petroleum, ½ oz. It is applied to the heads of the chicks in a semi-fluid state, and is said to be an infallible remedy. But the best cure is by means of the volatile powders named 'Kalydë,' made by Messrs. Chamberlin and Smith, of Norwich, or Spratt's 'Game Cure,' the introduction of which has made gapes no longer to be dreaded.

Indigestion is best known by the moping about of the birds, and on examination the breath is found to be very offensive. It is simply the result of feeding on too rich food, and if neglected develops into some more serious disease, so that it is important to take steps to overcome it at once. As soon as discovered put the birds on to very plain food, of which give very little, and mix a little powdered vegetable charcoal with it. By this means the birds will soon regain their normal condition, and then care must be taken to prevent a repetition of the same complaint.

Leg Weakness.—In adult fowls this trouble is usually due to excess of fat or straining to lay. The first thing is to give a pretty strong purgative—either a rhubarb pill or Epsom salts—which, by relieving the bowels, will most likely set up an altered condition of things. Repeat the dose after three days if there is any sign of constipation, and rub the legs and thighs well with turpentine or any good liniment.

Liver Disease.—Of this complaint there has been a great increase in recent years. It is generally due to bad feeding, and more especially to the use of Indian corn. The birds show dark or yellow face and comb, and gradually pine away. To get rid of the ill effects, give doses of Turkey rhubarb or any good liver medicine to all birds affected, on two alternate days, and then feed very sparingly until they have resumed their usual tone.

Prolapsus is a protrusion of the bowel, and is due to weakness of the organ, generally caused by over-stimulation. Return the bowel when it protrudes, and inject a weak solution of vinegar and water.

Rheumatism is sometimes found in old birds, and can be discerned by swelling of the limbs and heat therein. Of course it is evident that the bird is in pain. First give a mild aperient, and follow with doses of homœopathic tincture of aconite to allay the fever. Bathe the limbs in warm water, and after they have been carefully dried rub well with Elliman's or other good lotion, or turpentine, and encase in flannel. A little sulphur in the food will be beneficial.

Roup.—This complaint is a development of cold combined with tuberculosis. The reader will do well to refer to what is said when speaking of cold for treatment in the first stage, to which no more need be added. The difference between ordinary cold and roup is that when the latter is present the face and eyes swell, the nostrils become charged with a thick mucus, and the bird rapidly begins to waste away. Keep the bird in a warm place, and give roup pills, washing the eyes, head, and nostrils with a solution of permanganate of potash. Where worth the trouble and expense, a cure can generally be made, but it is most important to treat it in its first stage, before the disease has got a thorough hold upon the system, and when cured a fine day should be chosen upon which to return them to the runs, or they will have a relapse.

Scrofula.—Known by eruptions, and in any form is very difficult indeed to cure. Feed sparingly on wheat or barley, and give alternate with roup pills doses of some preparation of iron or quinine—say Parrish's Chemical Food.

Soft Crop arises from various causes, but generally is due to excessive thirst and a failure of the organ to perform its functions. When all the liquid cannot be got out of the crop by the mouth, puncture it with a coarse darning-needle, and so empty it. Place the fowl in a pen by itself, where it cannot obtain any food other than that supplied to it, and feed three times a day with very small quantities of soft food. A little water may be given, but only after each meal, and it should be acidulated with a little nitric acid. Give the bird homœopathic tincture of nux vomica twice a day, and keep penned for at least a fortnight.

Scaly Legs.—This trouble is due, in the first place, to a parasite burrowing under the scales, which causes them to rise, and ultimately fall off. It is most contagious, but not dangerous. Any bird affected should be removed to a pen by itself, and so kept until better. For treatment, wash the feet and legs with warm water and soap, rubbing the latter well in with a hard nail-brush, even if the scales fall off. Dry them perfectly, and anoint with carbolized vaseline, which can be obtained from any chemist, or one of the ointments specially prepared for the purpose. There is another form, due to dryness of the skin, and it may be treated in the same manner.

Soft Shells.—Shell-less eggs are due to a variety of causes. Perhaps the most common is a want of proper shell-forming materials, more especially when the hens are kept in confinement. As a preventive keep a supply in the run of old mortar, broken oyster-shells, coarse sand, or fine gravel, or other similar materials; common road-sweepings are also excellent for this purpose, as, in fact, is anything gritty which contains lime in one form or the other. Unslacked lime is, however, dangerous, and must not be used. Another frequent cause of hens laying shell-less eggs is that the egg organs are unduly stimulated, and the egg passes from the oviduct before it is coated with the shell. To remedy this, stop laying by feeding on hard food only, and give a dose of chlorodyne. This will give the organs rest, and allow nature to right itself.

White Comb.—Fowls in confinement often suffer from an unpleasant affection of the comb, which is covered with a white scurf. It is due to bad conditions, and these must be seen to. For external treatment use oxide of zinc and vaseline as an ointment.

Worms.—Fowls are often troubled with worms, which can be discerned in the excrement, and if these become excessive the birds cannot thrive. Keep them without food for a few hours, and then administer a dose of turpentine mixed with oil—say ten drops of turpentine to a teaspoonful of oil. After which keep them without food for six hours and administer a dose of castor oil.

CHAPTER XI

MARKETING THE PRODUCE

UNDER our complex systems of modern life it is necessary to realize that large masses of the population must depend upon others for almost all the food they consume, and some organization is absolutely necessary for meeting their needs. Upon the completeness or otherwise of the methods adopted to this end will depend the quality of produce when it reaches the customer, especially in perishable goods. There can be no question that food is better when it is fresh, ere it has lost those elements which go to make for flavour before decay sets in. Most of us have experienced the pleasure of eating various articles of food on the spot where grown or made, and they taste altogether different from those which have been kept for some time, perhaps knocked about in the hands of various traders. I know that imagination has a great influence upon our minds, but no one can really dispute the fact that all the more perishable food products should be marketed as rapidly as possible, even though we may not always be able to achieve this desirable state of things.

Eggs Perishable.—At one time it was difficult to make many people believe that eggs are very perishable, and speedily lose their fine quality. The fact that the contents of an egg are encased within a shell, that without breaking the hard covering we cannot test their freshness by smell, and that the eye, except to the few, does not help in this direction, has been the cause of want of appreciation. But nevertheless it is true that the fine quality of an egg is speedily lost, and unless marketed rapidly much of its value has disappeared. It is, further, very quickly affected by unfavourable conditions. The shell is porous, or permeable by air, and thus may be flavoured or tainted with those substances by which it is surrounded. At one time, and to some extent this is still so, Irish eggs were in great disfavour; being packed in damp or fusty straw, the entire lot were adversely affected thereby. An egg is one of the finest articles of food when good, but if tainted there is nothing more objectionable. Of course, as in other things, it is possible to retard the process of decay, for this has already set in when we find any sign of staleness present. The same is found in other foods, and modern science has done a great deal in this direction, but it cannot be claimed that preserved foods are equal to those which are fresh. To this end cold is found the best preservative, yet we have to pay a penalty for it. Frequently the method of preservation causes a change, as when milk is made into cheese and pork into bacon. With eggs in this country there is nothing of the kind.

Age of Eggs.—Millions of eggs are sold every year which cannot be regarded as really fit for food, though they are not decomposed enough to cause their condemnation. When we remember the distances they have to travel, coming from such far-off countries as Russia and Hungary, Italy and Canada, it will be seen that they must be weeks old ere reaching our shores. The people who eat them must be trained to do so. It is told of a Londoner who went to stay at a farmhouse that, when asked how he liked the new-laid eggs, he replied that they had not so much taste as those sold in London, which we can quite believe. We must expect that these long-distance foreign eggs will be stale, but it is unfortunately true that many of those produced at home are marketed when their best condition is past, due to the lethargic methods adopted, about which more is to be said. It must be our object to improve these methods, and when that is done no foreign egg can possibly compete with what is produced at home. Speaking generally, the limit of absolute freshness is when an egg is not more than three days old. Beyond that period it has passed the first stage, and ought never to be sold as new-laid, except it has been stored in a cold place, when the time may be extended for two or three days, if so kept from a few hours after it is laid.

Grades of Eggs.—Roughly speaking, eggs are divided into three classes, namely, (1) new-laid, (2) fresh, and (3) cookers. The last-named class are seldom sold by good retailers, but are employed for various manufacturing purposes, and find their way largely into low-class confectionery, where cheapness is the all-important consideration, so long as they are not absolutely bad. The first two classes often overlap each other. In some districts second-grade eggs are sold as new-laid, and in fact, as a rule, it is difficult to buy really prime eggs. These fresh eggs serve a useful purpose, and can be used for poaching, frying, or cooking, when, as the shell is broken, there is a certain evaportion of that which would be objectionable, if boiled, to those who know what a first-class egg really is. It should be the aim of all poultry-keepers to market the eggs when in the best condition, for the returns will be much higher than if at all doubtful in this respect. Every doubtful egg should be kept at home.

Testing Eggs.—Some first-class retailers never sell an egg unless it has been tested for freshness, finding that it is necessary to do so. The better plan would be for the vendors to make this test. If anyone gained a character for only selling eggs for what they really are, they would win confidence, and this must be a great gain. The simplest method is by light. If an egg is held in the hands between a strong light and the eye, the general condition can be discerned, and the size of the air-space noted. In a perfectly fresh egg this air-space is very small indeed, occupying about the space which could be covered by a sixpenny piece, at or near the broad end. As the egg advances in age evaporation takes place, and the air-space enlarges considerably, so that the period can to a large extent be determined by this test. Of course, the conditions under which the eggs are kept have much influence. If in dry warmth evaporation will be rapid; if in a cool place, slow. But the result is the same, for rapid evaporation means the same chemical change as would be seen by the slower process. A new-laid egg is milky in the white when boiled, whereas an older one becomes firmer and more glazy in the albumin. Apart from the mere idea of freshness, there can be no doubt that a new-laid egg is better as an article of food than one which is older. Housewives and cooks know that these really fresh eggs go much further than those that are stale, and we prefer the former even for poaching or frying. But, at the same time, second-grade eggs can be used for either of these purposes without any fear, and will not be as objectionable as they might be if boiled in the shell.

Packing Eggs.—A most important question to those who forward eggs to retailers or others is how to pack them in the best, cheapest, and safest manner. Where very large quantities are forwarded, as in the case of collectors, wooden boxes or baskets, lined with clean straw or wood wool, and with the same material between each row, are generally preferred; but enough care is not usually taken in the packing. It is desirable that the eggs shall be clean, evenly laid, and whilst firm, yet not crushed. We have yet much to learn in this direction, and the National Poultry Organization Society has been formed to secure a better system of packing and greater rapidity in marketing eggs, and to encourage the general development of the poultry industry. Where eggs are sent a good distance it is better to use non-returnable cases, saving all the carriage of empties and the trouble arising therefrom. Foreign eggs are often preferred by traders for this reason. Where smaller quantities are handled, it is better to use the patent egg-boxes now being sold by various makers. Of these we need only mention three. The oldest is the box sold by the Dairy Supply Company of London. These have ordinary square cardboard partitions, with felting between each row, and they carry very well indeed. The Pocock egg-boxes (Fig. 18), made by the Dairy Outfit Company of London, differ in essential principles, in that the cases are fitted with trays standing one on another. The trays have rods across, on which is laid felting, and thus the eggs are safe against concussion. We have known one of these boxes used regularly for months without a single breakage, and they are very extensively employed. Recently Mr. J. Robinson, of Clitheroe, has brought out a cardboard-partitioned egg-box called the Honeycomb (Fig. 19), which gives great economy of space. In this the partitions are made like a honeycomb, so that all corners are avoided, and there is great economy of space.

Sorting.—Eggs should always be sorted to about the same size, extra large or very small eggs being kept at home. I hope that we shall yet see eggs sold by weight, but in the meantime it is wise to send all of the same size in a box. Care should be taken to send out eggs clean, and it is better to wash them if this is necessary; but we prefer to see the bloom on the shell, which would be removed if washed.

Table Poultry.—In all industries the ideal would be that producers and consumers should come into direct contact with each other, and in former days this was the case to a greater extent than now. But with the growth of great centres of population and rapid intercommunication between all parts of the country this can only be to a limited extent. Whenever possible—and this ought to be so to a greater degree than is generally supposed—direct sales should be made, to which end markets should be encouraged. But however much we may declaim against the middleman, he is a necessary factor; and so long as he is kept in his right position—that is, as an intermediary between one class and another, and not a dominator—he serves a most useful purpose, earning well the commission allowed him, and saving more than his cost. But if the result is at once to depreciate the return obtained by producers and enhance the cost to consumers, then we have a very serious state of things. In practice we know that better prices can be obtained for the great bulk of table poultry from salesmen and dealers than in any other way, more especially by those who have good birds to sell. Complaints are chiefly rife amongst those who send forward mediocre or poor qualities of any produce. We are bound to recognise, therefore, that it is to 'the trade' that we must look for help in this direction, and the greatest success will be achieved by producing for it, unless there is a good local demand retail. When fowls are sold to retailers, they should only be killed, plucked, and shaped, full particulars of which are given in 'Poultry-Keeping as an Industry for Farmers and Cottagers,' and not drawn or trussed; but where sold direct to consumers, it is better to prepare ready for cooking, as they will thus command a much better price.

INDEX

THE END

www.ingramcontent.com/pod-product-compliance
Ingram Content Group UK Ltd.
Pitfield, Milton Keynes, MK11 3LW, UK
UKHW020843290926
15470UKWH00004B/1517

9 789349 690875